Genesis and Development of a Scientific Fact

*Edited by Thaddeus J. Trenn
and Robert K. Merton*

*Translated by Fred Bradley
and Thaddeus J. Trenn*

Foreword by Thomas S. Kuhn

Genesis and Development of a Scientific Fact

Ludwik Fleck

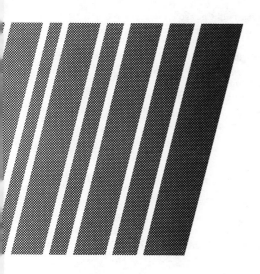

The University of Chicago Press

Chicago and London

Originally published as *Entstehung und Entwicklung einer wissenschaftlichen Tatsache: Einführung in die Lehre vom Denkstil und Denkkollektiv.* © 1935 by Benno Schwabe & Co., Basel, Switzerland.

The University of Chicago Press, Chicago 60637
The University of Chicago Press, Ltd., London

Library of Congress Cataloging in Publication Data

Fleck, Ludwik.
  Genesis and development of a scientific fact.

  Translation of Entstehung und Entwicklung einer
wissenschaftlichen Tatsache.
  Bibliography: p.
  Includes index.
  1. Science—Social aspects. 2. Science—Philosophy.
3. Syphilis—Diagnosis—Wassermann reaction.
4. Knowledge, Sociology of. I. Title.
Q175.5.F5913   507'.2   79-12521
ISBN 0-226-25324-4
ISBN 0-226-25325-2 (paper)

Contents

Contents

Illustrations

Foreword

The appearance of an English translation of Ludwik Fleck's *Genesis and Development of a Scientific Fact* realizes a project I have urged on various friends and acquaintances (but not on the present editors) ever since I first encountered the book a quarter of a century ago. My purpose in calling for a translation was not simply to make Fleck's work accessible to an English-speaking audience but rather to provide it with an audience at all. In twenty-six years I have encountered only two people who had read the book independent of my intervention. (One was Edward Shils, who has apparently read everything; the other Mark Kac, who had known the author personally.) The editors of this edition tell me that they too first learned of the book from me. Under the circumstances, I have not known how to decline their invitation to provide a prefatory note reflecting on my own encounter with Fleck.

To the best of my recollection, I first read the book during the year 1949 or early in 1950. At that time I was a member of the Harvard Society of Fellows, trying both to prepare myself for the transition from research physics to history of science and also, simultaneously, to explore a revelation that had come to me two or three years before.[1] That revelation was the role played in scientific development by the occasional noncumulative episodes that I have since labeled scientific revolutions. On that non-topic there was no established bibliography, and my reading was therefore exploratory, often owing much to serendipity. A footnote in R. K. Merton's *Science, Technology, and Society in Seventeenth Century England*[2] led me to the work of the developmental psychologist Jean Piaget. Though Merton's book was an obvious desideratum for a prospective historian of science, Piaget's work surely was not. Even more unlikely was the footnote that led me to Fleck. I found it in Hans Reichenbach's *Experience and Prediction.*[3]

Reichenbach was not, of course, a philosopher who thought of facts as possessing a life cycle. Citing the plate in which Fleck displays changing representations of the human skeleton, he wrote: "Intellectual operations have shown us the way to overcome the limitations of our subjective intuitional capacities.... Every picture may, besides containing false traits, introduce some true features into the composition."[4] Fleck would no more have written those sentences than Reichenbach would have spoken of the "genesis and development of a scientific fact." But the latter was Fleck's title, and Reichenbach had to include it when citing the plate. Reading the citation, I immediately recognized that a book with that title was likely to speak to my own concerns. Acquaintance with Fleck's text soon confirmed that intuition and inaugurated my never-very-systematic campaign to introduce a wider audience to it. One of those to whom I showed it was James Bryant Conant, then President of Harvard but shortly to become the U.S. High Commissioner for Germany. A few years later he reported with glee the reaction of a German associate to his mention of the title: "How can such a book be? A fact is a fact. It has neither genesis nor development." That paradox was, of course, what had drawn me to the book.

I have more than once been asked what I took from Fleck and can only respond that I am almost totally uncertain. Surely I was reassured by the existence of his book, a nontrivial contribution because in 1950 and for some years thereafter I knew of no one else who saw in the history of science what I was myself finding there. Very probably also, acquaintance with Fleck's text helped me to realize that the problems which concerned me had a fundamentally sociological dimension. That, in any case, is the connection in which I cited his book in my *Structure of Scientific Revolutions.* [5] But I am not sure that I took anything much more concrete from Fleck's work, though I obviously may and undoubtedly should have. At the time I found Fleck's German extraordinarily difficult, partly because mine was rusty and partly because I possessed neither the background nor the vocabulary to assimilate discussions of medicine and biochemistry, especially when viewed from the to me unknown and yet vaguely repulsive perspective of a sociology of the collective mind. The lines in the margin of my copy of the book suggest that I responded primarily to what had already been very much on my mind: changes in the gestalts in which nature presented itself, and the resulting difficulties in rendering "fact" independent of "point of view." Even at that date, though much engaged in exploring Kohler, Koffka, and other gestalt psychologists, I resisted (as Fleck surely would have) their frequent substitution of "seeing as" for "seeing." What I saw when I looked at the famous duck-rabbit was either the duck or the rabbit, but not the lines on the page—at least not until after much conscious effort. The lines were not the facts of which the duck and rabbit were alternative interpretations.

Rereading the book now, as I have not done in the interim, I find many insights that I might fruitfully have worked into my viewpoint. I am, for example, much impressed by Fleck's discussion (chap. 4, sec. 4) of the relation between journal science and vademecum science. The latter may conceivably be the point of origin for my own remarks about textbook science, but Fleck is concerned with another set of points—the personal, tentative, and incoherent character of journal science together with the essential and creative act of the individuals who add order and authority by selective systematization within a vademecum. Those issues, which had escaped me entirely, merit much additional consideration, not least because they can be approached empirically. Again, given my own special concerns, I am particularly excited by Fleck's remarks (chap. 4, sec. 3) on the difficulties of transmitting ideas between two "thought collectives," above all by the closing paragraph on the possibilities and limitations of participation in several "thought communities." ("Very different thought styles are used for one and the same problem more often than are very closely related ones. It happens more frequently that a physician simultaneously pursues studies of a disease from a clinical-medical or bacteriological viewpoint together with that of the history of civilization, than from a clinical-medical or bacteriological one together with a purely chemical one.") Here, too, Fleck opens avenues for empirical research.

Readers will find many other similar *aperçus* in Fleck's rich and penetrating book. Though much has occurred since its publication, it remains a brilliant and largely unexploited resource. But the position it elaborates is not free of fundamental problems, and for me these cluster, as they did on first reading, around the notion of a thought collective. What troubles me is not that a thought collective is a hypostatized fiction, though I think it is. Fleck's own response to *that* objection (chap. 4, sec. 3, n. 7) seems adequate. Rather I find the notion intrinsically misleading and a source of recurrent tensions in Fleck's text.

Put briefly, a thought collective seems to function as an individual mind writ large because many people possess it (or are possessed by it). To explain its apparent legislative authority, Fleck therefore repeatedly resorts to terms borrowed from discourse about individuals. Sometimes he writes of the "*tenacity* of closed systems of opinion" (chap. 2, sec. 3; italics mine).[6] Elsewhere he accounts for this tenacity in terms, for example, of "trust in the initiated, their dependence upon public opinion, intellectual solidarity" (chap. 4, sec. 3). Responding to these forces, the members of a successful thought collective come to participate in what Fleck sometimes describes as "a kind of harmony of illusions" (chap. 2, sec. 3). Doubtless that last phrase is intended metaphorically, but it is a damaging metaphor, for it reinforces the impression that, in the absence of social pressures, illusion might have been avoided: "If [Siegel's] findings had had the *appropriate* influence and received a *proper* measure of publicity..., the concept of syphilis would be different today" (chap. 2, sec. 4; italics mine).

Other phrases throughout Fleck's book suggest a very different position, one far closer to my own. The force of a thought collective is, for example, sometimes described as a "compulsion" or an "intrinsic constraint" (chap. 3). Elsewhere Fleck writes that "the communication of thoughts within a collective, irrespective of content or logical justification, should lead for sociological reasons to the corroboration of the thought structure [*Denkgebilde*]" (chap. 4, sec. 3). These and many other phrases in the book indicate that the effects of participation in a thought collective are somehow categorical or a priori. What the thought collective supplies its members is somehow like the Kantian categories, prerequisite to any thought at all. The authority of a thought collective is thus more nearly logical than social, yet it exists for the individual only by virtue of his induction into a group.

That position is, of course, extraordinarily problematic, and Fleck's occasional attempts to develop it by distinguishing between the passive and active elements of knowledge are to me unenlightening. "Passive" and "active" are again terms borrowed from individual psychology for application to a collective. Far more useful in these and other passages would be the epistemic distinction between knowledge and belief. (For example, if "knowledge" were substituted for "opinion" in the previously quoted phrase—"the tenacity of self-contained systems of opinion"—then "tenacity" and perhaps also "self-contained" and "systems" would become redundant.) Yet, to note these central problems is scarcely to discredit Fleck. The difficulties to which these closing paragraphs point have been central to philosophy since Wittgenstein,[7] and they remain unsolved. To have disclosed them in the empirical materials of the history of science may be a sufficient achievement.

Thomas S. Kuhn
June 1976

Preface

In 1974 I gave a series of seminars at the University of Regensburg on the work of Thomas S. Kuhn, including that of his predecessors and contemporaries. During the winter semester we analyzed Ludwik Fleck's 1935 monograph *Entstehung und Entwicklung einer wissenschaftlichen Tatsache*. We found among other things that Fleck carefully distinguishes paradigm from thought style, the former as exemplar and the latter as that which sociologically conditions cognition within the thought collective. The conceptual creations of science, like other works of the mind, become accepted as fact through a complex process of social consolidation. These thought products, and the thought style under which they arise, are never finalized but can undergo transformation through intracollective or even intercollective interaction, whereby symmetry is democratically preserved between the esoteric circle of experts and the exoteric circle of the wider society, and marginal men participating in diverse thought collectives can create something new from the conflict. The older way of looking at things may become incomprehensible under the new thought style, and the process of transformation from one to the other may be a rapid gestalt switch or a slow process of differentiation like that between variation and species.

Fleck is able to discuss scientific change without dwelling upon the controversial concept of revolutions; indeed, Kuhn's sharp distinction between so-called normal and revolutionary science has no direct counterpart in Fleck's theory. We were also struck by his insistence upon the method of comparative epistemology, which is becoming increasingly recognized for its importance in the history of science.[1] Fleck develops his central ideas not merely in the abstract but through a beautifully analyzed historical case study of syphilis and the pathbreaking discovery of the Wassermann reaction, accepted only because it proved to be extremely useful.

The relevance of Fleck for current research in the sociology, history, and philosophy of science was so overwhelming that I decided to bring out an English-language edition of his book. I soon realized that to translate this difficult text alone, with its highly idiosyncratic style, would be far too time-consuming and beyond the capabilities of any one individual. A solution would be to have an initial translation by an experienced professional translator, which could then be appropriately modified by terminological and interpretive editing. But translation costs were prohibitive and no grants were forthcoming. Reluctant to abandon what seemed a very worthwhile project, I then sought the aid and advice of Robert K. Merton. He had known of Fleck's book only through Kuhn's early allusion to it but, upon reading the monograph, agreed that it was long overdue for translation into English. He proceeded to arrange for publication with the University of Chicago Press, drew upon the research grant by the National Science Foundation to the Columbia University Program in the Sociology of Science for the funds required in support of the initial translation, formally invited Thomas S. Kuhn to contribute a Foreword, and agreed to serve as coeditor of the volume. Few books can be more appropriate than Fleck's for an international rebirth. Written in German by a Jewish physician-scholar from Poland, first published in Switzerland, it has been translated in Oxford and Regensburg, edited in New York and Regensburg, and appears under a Chicago imprint. In effect, this translated edition is itself a collective outcome of the very type Fleck describes. The currently intensified interest in the interaction between the sociology, history, and philosophy of science makes this pioneering work more timely now than when it first appeared more than forty years ago.

The translation was a difficult task. Fred Bradley in Oxford, who has translated over thirty-five books, accepted this project as a unique challenge. Merton and I agreed that a close and literal translation of the complex and often idiosyncratic original would only result in an unreadable English version. Bradley's initial translation was largely reworked by the editors from the point of view of interpretation, terminology, and style. In the interests of preserving accuracy and conveying meaning, this edited version departed at times from the German original, and so was resubmitted to Bradley for further discussion and compromise where appropriate. The final result is an interpretative translation, which seeks to remain faithful to Fleck's complex ideas while conveying them in a readable fashion.

A word must be added about some of the key terms. Some words quite resist adequate translation, and perhaps none more than the central terms *Denkstil* and *Denkkollektiv*. One might have justified the retention of *Denkstil* following the precedent of *Jugendstil*, but we preferred to provide a translation. For the German *Denkstil* it seemed best to adopt "thought style" as a straightforward equivalent, although it transmits few of the cultural overtones inherent in the original and places undue emphasis upon presumed rational processes. The German word *Denkstil* was introduced by Karl Mannheim in 1925, and remains in current use.[2]

We were less inclined initially to preserve the original *Denkkollektiv*, since "thought collective" seemed awkward beyond ready acceptance. The term *Denkkollektiv* appears to have been introduced by Fleck and is not standard German usage. We considered introducing a neologism, but a close editing of the translation indicated the necessity of retaining "thought collective" if the conceptual balance of the text was to be preserved. Fleck uses the adjective *kollektiv* and the noun *Kollektiv* in a great variety of related contexts throughout the text, and it would have been unjustified not to translate this with the intended sociological term "collective." Since *Denkkollektiv* refers to that same collective which engages in *kollektive Arbeit*, has a *kollektiven Denkstil*, produces *kollektive Erfahrung*, has both *kollektive Gedanken* and *kollektive Vorstellungen*, operates according to *kollektiven Denklinien*, and produces *kollektive Gebilde*, we have designated it simply the "thought collective" in order to maintain unity throughout the work. The various and rather different senses in which Fleck uses *Denkkollektiv* will thus not be assigned different English terms as was initially considered but will in each case be translated as "thought collective." But for the fact that Fleck explicitly meant "collective" rather than "community"—a highly contentious terminological issue—we might have been able to adopt the symmetry of "style of thought" and "community of thought." The "thought collective of modern science" (*moderne wissenschaftliche Denkkollektiv*) could then have been conveniently rendered as the "modern scientific community." But we have refrained from such license throughout the translation.

It should be noted that we entertained the possibility of using "school of thought" for *Denkstil* and/or *Denkkollektiv*. But "school of thought" is especially ambiguous in the context of Fleck's ideas, carrying aspects of both *Denkstil* and *Denkkollektiv*. It is thus not appropriate in a context where these must be considered distinctive even though they are inseparable one from the other. Utilization of the familiar "cognitive style" and "cognitive community" based on the Latin *cognoscere*, would be misleading, since both *Denkstil* and *Denkkollektiv* are based on the allied but different *cogitare*.

There are other difficult terms. The word *Lustseuche* as used by Fleck cannot always be rendered as "syphilis" but must often retain the crucial element of punitive plague or scourge resulting from lecherous and sinful fornication. While "lecher's-disease plague," or "coital plague" were considered, the meaning seems best carried by "carnal plague or scourge." Thus, depending on the context, *Lustseuche* has been rendered variously by "the carnal scourge," "lues venerea," or "the great pox," where the latter term balances nicely with "the French pox" as well as with the non-venereal variola "small pox" and "swine pox" in certain portions of the text. A similar difficulty arose in the case of *Sinn-Bild* and *Sinn-Sehen*. At one point Fleck identifies *Sinn-Bild* with the technical term "ideogram," so that we adopt the neologism "ideovision" for *Sinn-Sehen*—the faculty to apprehend, visually perceive, or formulate ideograms. Such terms as "collective experience" have sometimes been given in the forms "experience of the collective," "mood of the collective," or "effort of the collective" to emphasize the sense conveyed by the original. The "thought style of modern science" seems preferable to the "modern scientific thought style" so as not to overload the term with the significance of a general cultural trait. The frequent appearance of "mood" for *Stimmung* is generally interchangeable with "disposition" throughout the text; "temper" or even "temperament" could have been used in most contexts. Other terms that are of sufficient difficulty or interest are noted editorially as they arise.

Fleck's pioneering monograph was published at almost the same time as Karl Popper's *Logik der Forschung*. But, developed in very different cognitive styles, the books met with contrasting response. In Popper's own words, his book "was surprisingly successful, far beyond Vienna. There were more reviews, in more languages, than there were twenty-five years later of [its translation] *The Logic of Scientific Discovery*, and fuller reviews even in English" (*The Philosophy of Karl Popper*, ed. P. A. Schilpp, La Salle, Ill., 1974, 1:85). But, as we have seen, Fleck's book had negligible influence. The dominant thought style of the 1930s was not one in which Fleck's seemingly idiosyncratic ideas would resonate widely. Much the same fate attended another monograph in the sociology of science appearing three years after Fleck's work—Robert Merton's *Science, Technology, and Society in Seventeenth-Century England*—which also experienced a substantial delay of significant response by the community of sociologists and historians of science. Such delays in the impact of scholarly works invite speculative interpretation. In the case of Fleck, it may have been the case that his analysis of the history of syphilis and of the discovery of the Wassermann reaction was primarily perceived as another historical case study in medicine and could stand on its own grounds as such. Our search has yielded nearly a dozen reviews at the time, most of them in medical journals and most in German. These did not uniformly note that, rather than being only an exercise in the history of medicine, the monograph was for Fleck primarily a vehicle for his intricate theory intertwining the sociology, history, and philosophy of science. It is perhaps most diagnostic that the book received no review notice at all in George Sarton's *Isis,* by then the leading international journal of the history of science.

Fleck may have been disappointed but was surely not surprised at the modest reception accorded his ideas. The time was not yet ripe and he knew it. With remarkable prescience, however, he forecast in 1935 the emergence of a thought style compatible with his own theory of thought style and thought collective. "The thought style is a social product: it is formed within a collective as the result of social forces. This circumstance links problems of natural science with those of sociology and especially the sociology of thought, a science which unfortunately still remains grossly neglected, but is bound soon to move into the center of intellectual interest."[3]

The monograph was published in an edition of 640 copies, of which about 200 were sold. The *National Union Catalog* lists six copies in major libraries in the United States but does not mention the one at Harvard. Fleck receives no notice at all in a sampling of standard works in the history of medicine and the sociology of knowledge. Indeed, the first published reference to his monograph since the 1930s seems to be that of Kuhn in 1962.

Of late this situation has changed dramatically. In 1976, I announced that the present edition was forthcoming.[4] Quite independently of this translation-edition, W. Baldamus in Birmingham, England, also became interested in Ludwik Fleck. He has already published portions of his own English translation, offering an alternative approach to Fleck's work.[5] As our own, authorized edition appears, then, there is ample evidence that Fleck's day has come,[6] that the elucidation of his central ideas is gathering force.

True to the spirit of Fleck, this edition is a collective effort; a social product with a variety of contributions both esoteric and exoteric over a wide range of fields of interest. I am grateful to each and every one who helped, including Margaret Bradfield, Fred Bradley. Rosemarie Buchmayer, Claude Dolman, Ernestina Fleck, Erika Hickel, Bernd Janele, Gundolf Keil, Marcus Klingberg, Hugo Kuhn, Thomas Kuhn, Wolfgang Lenzen, Erna Lesky, Nikolaus Mani, Robert Olby, John Parascandola, Christoph Puschmann, Nora Sigerist-Beeson, Glenn Sonnedecker, and Reiner Wieland.

Thaddeus J. Trenn

Genesis and Development of a
Scientific Fact

Overview of Contents

Prologue

What is a fact?

A fact is supposed to be distinguished from transient theories as something definite, permanent, and independent of any subjective interpretation by the scientist. It is that which the various scientific disciplines aim at. The critique of the methods used to establish it constitutes the subject matter of epistemology.

Epistemology often commits a fundamental error: almost exclusively it regards well-established facts of everyday life, or those of classical physics, as the only ones that are reliable and worthy of investigation. Valuation based upon such an investigation is inherently naive, with the result that only superficial data are obtained.

Moreover, we have even lost any critical insight we may once have had into the organic basis of perception, taking for granted the basic fact that a normal person has two eyes. We have nearly ceased to consider this as even knowledge at all and are no longer conscious of our own participation in perception. Instead, we feel a complete passivity in the face of a power that is independent of us; a power we call "existence" or "reality." In this respect we behave like someone who daily performs ritual or habitual actions mechanically. These are no longer voluntary activities, but ones which we feel compelled to perform to the exclusion of others. A better analogy perhaps is the behavior of a person taking part in a mass movement. Consider, for instance, a casual visitor to the Stock Exchange, who feels the panic selling in a bear market as only an external force existing in reality. He is completely unaware of his own excitement in the throng and hence does not realize how much he may be contributing to the general state.* Long-established facts of everyday life, then, do not lend themselves to epistemological investigation.

As for the facts of classical physics, here too we are handicapped by being accustomed to them in practice and by the facts themselves being well worn theoretically. I therefore believe that a "more recent fact," discovered not in the remote past and not yet exhausted for epistemological purposes, will conform best to the principles of unbiased investigation. A medical fact, the importance and applicability of which cannot be denied, is particularly suitable, because it also appears to be very rewarding historically and phenomenologically. I have therefore selected one of the best established medical facts: the fact that the so-called Wassermann reaction is related to syphilis.

HOW, THEN, DID THIS EMPIRICAL FACT ORIGINATE AND IN WHAT DOES IT CONSIST?

Lvov, Poland, summer 1934

---

*"Those who join a panic make a panic." H. G. Wells, 1916.—Eds.

| One | # How the Modern Concept of Syphilis Originated |

The historical sources of syphilology can be traced back, without a break, to the end of the fifteenth century. They contain descriptions of a more or less differentiated specific disease (in modern terms a so-called *disease entity*) which historically corresponds to our concept of syphilis, although the bounds and nomenclature have undergone considerable modification. The symptomatology of the disease also underwent a similar transformation. Toward the end of the fifteenth century the line of development in the diagnosis of syphilis disappears from our view into an undifferentiated and confused mass of information about chronic diseases characterized by skin symptoms frequently localized in the genitals—diseases that sometimes assumed epidemic proportions.

Within this primitive jumble of the most diverse diseases, which crystallized during the following centuries into various entities, we can detect in addition to syphilis what we now call leprosy; scabies; tuberculosis of the skin, bone, and glands; small pox (variola); mycoses of the skin; gonorrhea, soft chancre, probably also lymphogranuloma inguinale, and many skin diseases still regarded as nonspecific today, as well as general constitutional illnesses such as gout.

The confused political conditions prevailing in Europe at the close of the fifteenth century with wars, famine and natural disasters, such as extreme heat and floods affecting many localities, resulted in a dreadful accumulation of divers epidemics and diseases.[1] These occurred with such frequency and brought such fearful misery that the attention of research workers was increased, initiating the development of syphilological thought.

One particular circumstance above all others, namely the astrological constellation, if not father to this thought at least sired one of its constituents. "Most authors assume that the conjunction of Saturn and Jupiter under the sign of Scorpio and the House of Mars on 25.XI.1484 was the cause of the carnal scourge [*Lustseuche*]. Benign Jupiter was vanquished by the evil planets Saturn and Mars. The sign of Scorpio, which rules the genitals, explains why the genitals were the first place to be attacked by the new disease."[2]

Astrology played a dominant role at the time, which readily explains why the astrological interpretation of the origin of syphilis had such a persuasive effect upon then-current research. One also finds that almost all the authors of that period hint at the sidereal origin of syphilis as the first and most important cause of the epidemic. "Furthermore, the condition here mostly affects the genitals initially, spreading from there to the whole body, *and no other disease is found that starts in this way.* But I guess that this is brought about by some affinity between the genitals and this disease. This may derive from some celestial effect, as the astrologers claim, arising from the conjuction of Saturn and Jupiter in the third aspect of Scorpio in the 23d degree in 1484 as well as from a simultaneous configuration of other fixed stars which just happened to occur then. Over long stretches of time many diseases were seen to arise, as well as old ones to die out, as we shall clearly show later. Not only is the origin of this disease traced to the position of the stars, but the disease is fomented again and again especially by the sign of Scorpio, which rules the genitals."[3]

The explanation given to any relation can survive and develop within a given society only if this explanation is stylized in conformity with the prevailing thought style. Astrology thus contributed its share to the firm establishment of the venereal character of syphilis as its first "specific difference." Religious teachings, claiming that the disease is a punishment for sinful lust and that sexual intercourse has special ethical significance, finally established this cornerstone of syphilology, ascribing to it a pronounced ethical character. "Some refer the cause of the disease to God, Who has sent it because He wants Mankind to shun the sin of fornication." [4]

If the epidemic provided the subject matter for an investigation, the resulting emergency stimulated it. Astrology was the dominant science, and religion created a mystical frame of mind. Together these produced that sociopsychological prevailing attitude which for centuries favored the isolation and consistent fixation upon the emotive venereal character of this newly determined disease entity. The stigma of fatefulness and sinfulness was imprinted upon syphilis—a stigma which it still carries within large sectors of the public.

This rudimentary idea in syphilology, involving a theory of the venereal nature of syphilis, [5] or syphilis construed simply as the carnal scourge, seems far too broadly based. It comprises not only what we today call syphilis but also the other venereal diseases, which have so far been successively distinguished as gonorrhea, soft chancre, and lymphogranuloma inguinale. However, the sociopsychological and historical foundation was so strong that it took four centuries before scientific advances in other fields were important enough to establish a definitive distinction among these various diseases. Such entrenchment of thought proves that it was not so-called empirical observations that led to the construction and fixation of the idea. Instead, special factors of deep psychological and traditional significance greatly contributed to it.

This first feature in the emergent science of syphilology began to establish itself toward the end of the fifteenth and during the course of the sixteenth century. But it did not remain the only one. Three other ideas, originating in other social strata and during other eras, interacted with it. Only through this interaction, the cooperation and opposition among these ideas, has the definition of syphilis as a disease entity been advanced to its present level.

The second idea evolved from medical practitioners using pharmaceutical resources. Sudhoff comments: "As a result of decades of practice, certainly spanning several generations, it became possible to distinguish and isolate from the host of chronic skin conditions a particular group which, when treated with mercury ointment, reacted favorably, indeed was completely cured. . . . This therapeutic finding was also taken up by the internal specialists. During the middle of the fourteenth century we encounter for the first time a comprehensive designation of those chronic skin diseases that can be cured by treatment with this general mercurial embrocation. These were distinguished from the many variants of scabies, namely chronic eczema and related skin diseases such as scabies grossa."[6]

Sudhoff regards the use of mercury, which is rooted in very ancient metallotherapy, as the true and only origin of the syphilis concept. But this theory seems to me to be incorrect. Some early treatises on the subject consider syphilis a disease entity but do not even mention mercury. Second, mercury was a popular remedy for many other diseases of the skin such as scabies and leprosy. Third, if the curative effect of mercury were alone decisive, other venereal diseases such as gonorrhea and soft chancre should not have become related to syphilis at all, since these remain unaffected by mercury. The curative effect of Hg seems therefore to be only a secondary factor in the establishment of the concept of syphilis.

Its importance must nevertheless not be underrated, for the use of mercury in the treatment of syphilis was very widespread. It is said, for instance, that "metallic remedies are mainly mercury" or that a particular "substance is combined with these metals, mostly with mercury, although I use cinnabar more than sublimate."[7] Remarkably, even the flow of toxic saliva during mercury treatment was considered to be a therapeutic effect involving the "evacuation" of the syphilis toxin. Such an evacuation occurs "mainly through the sputum, and nothing is more efficacious than mercury in promoting it."[8]

The use of mercury in treating syphilis has been traditionally regarded as natural. Although it presented the risk of poisoning, it was nevertheless felt that "mercury is noble, useful in many fields, and necessary."[9] As time went on, the effect of Hg became increasingly recognized and mercury was also used as a diagnostic aid.

But even as late as the nineteenth century it was still not possible to settle upon a satisfactory concept of syphilis on the basis of Hg

alone. In accordance with the idea of carnal scourge, syphilis was
thought to include other venereal diseases such as gonorrhea and
soft chancre and their complications. These, as well as local diseases
of the genitals, such as balanitis, which are regarded as nonspecific
even today, were later differentiated on the basis of pathogenesis
and etiology. These latter conditions, however, remain unaffected
by mercury. So to combine both points of view, that of mercury
and that of carnal scourge, it was observed that "sometimes mercury does not cure the carnal scourge but makes it even worse."[10]
This neatly resolved the dilemma. The mercury idea really concerned the diagnosis only of so-called constitutional syphilis, that
is, the stage of the generalized disease. The primary stage, properly
venereal because it is localized in the genitals, remained untouched
by that idea and was characterized by the idea of the carnal scourge.

Thus two points of view developed side by side, together, often at
odds with each other: (1) an ethical-mystical disease entity of
"carnal scourge," and (2) an empirical-therapeutic disease entity.
Neither of these points of view was adhered to consistently. Although mutually contradictory, they eventually became amalgam-
ated. Theoretical and practical elements, the a priori and the
purely empirical, mingled with one another according to the rules
not of logic but of psychology. Empiricism was largely displaced by
an emotive apriority. Some physicians even doubted the existence
of syphilis altogether. In a sixteenth-century treatise it is claimed
that "quite a few therefore state that there is no such thing as the
French pox, except in the imagination of some of our contemporaries. For they say that what we call the French pox constitutes
a variety of conditions."[11]

There were those who doubted it even as late as the end of the
nineteenth century. Dr. Josef Hermann, for many years (1858–88)
physician-in-chief and head of the department of syphilis at the
Imperial and Royal Hospital of Wieden in Vienna, published a
pamphlet about 1890 entitled *Constitutional Syphilis Does Not
Exist.*[12] In his view syphilis is a "simple, local disease which never
spreads to the human blood, is completely curable, never leaves
permanent effects, and is never propagated by procreation and
heredity." It manifests itself through chancre or gonorrhea "and
by all the direct sequelae of these two primitive conditions." On the

other hand, there is a "veritable host of disease manifestations which profoundly affect the social life of humanity and even the entire generation. But all these general symptoms do not constitute syphilis at all. They are exclusively produced either by the mercury treatment itself or by other bad concoctions." To Hermann, syphilis still meant the original carnal scourge [*Lustseuche*], construed as only a localized disease without general symptoms. A general syphilitic condition would have the presence of syphilis in the blood as a "first premise," but "the existence of syphilitic blood is only a dogmatic precept without the slightest evidence to support it." Hermann declared that "no pathognomonic indication of syphilis will ever be found in the blood of those suffering from this disease."

Although his views appear to have been fossilized as measured even by the standards of his time, this outsider is of particular importance in our investigation for the following reason. He attested to just how strongly syphilis and Hg had become associated with one another, and also emphasized the quandary in which the physicians found themselves because of the pleomorphism of the syphilis symptoms. This in turn produced a general and urgent "demand for blood tests" as the means to identify this disease entity with precision.[13]

The concept of syphilis, then, was still vague and incomplete. The two approaches to it contradicted each other. The clash was felt all the more clearly to the extent that the primitive ethical-mystical idea lost its fascination owing to changes in the general thought style and as more details of the relevant phenomena were brought to light. But the concept still remained too variable, and not sufficiently entwined or interwoven within the fabric of contemporary knowledge to be considered finally established with a firmly based, objective existence, and thus to appear as an undoubtedly "real fact."

The intellectual clarity of the issue, in particular, was clouded because several important factors were disregarded. No attempt was made to differentiate between venereal diseases with general symptoms and those either lacking them altogether or, like gonorrhea, rarely exhibiting them. Hereditary syphilis and the inferior quality of the offspring of syphilitic parents were also problems; as were the unsolved puzzle of latent syphilis, the reappearance of the disease, and also its variously suspected relation to many other diseases such as tabes and progressive paralysis, lupus, and scrophulosis. In general, the era of experiments and wide-ranging knowledge developed in detail had already begun. History records an abundance of experiments and observations about inoculations, reinoculations, and immunity relations. But it would be wrong to think that experiments, no matter how clearly conceived, always produced the "correct" results. Although important as starting points for new methods, these were worthless as evidence.

An argument developed between those who believed in the identity of gonorrhea, syphilis, and soft chancre (the "theory of identity") and the physicians who wanted to divide the great pox into several disease entities. "Several physicians, especially Andree and Swediauer, attempted to establish the identity of the infective material of the two diseases on the basis of the gonorrheic mucus and the chancre pus. After a few experiments conducted with these substances it was claimed that the gonorrheic toxin could sometimes produce chancre and vice versa—a view which was widely adopted. Fritze thought that the two conditions differed specifically but not generically,"[14] since with some organisms the pathogenic substance "was too weak to produce chancre, although still strong enough to produce gonorrhea." Hunter[15] inoculated the skin of a healthy person's genitals with gonorrheic pus and obtained an ulcer followed by typical syphilis. Although he maintained the identity of gonorrhea with syphilis, he differentiated between soft and hard (indurated) chancre; the latter alone was supposed to be part of the syphilis complex (the "theory of duality"). A theory of pseudosyphilis grew out of the distinction. This involved a disease resembling syphilis, although it was in fact fundamentally different and not preceded by hard chancre.

Another school differentiated the gonorrheic toxin from that of syphilis, but regarded gonorrhea as the primary stage of a general constitutional disease designated the "gonorrhea disease." This indicates the influence of the theory of syphilis. The "unitarians" (Ricord),[16] yet another school at the time, completely separated gonorrhea from syphilis. They insisted, however, that both hard and soft chancre were identical and spoke of a special predisposition to general syphilitic afflictions, necessary for the general stage to succeed chancre. Finally, the "new theory of duality"[17] clearly differentiated both gonorrhea and soft chancre from syphilis.

All these observations refer only to the question how to distinguish among various venereal diseases. By no means does it concern the whole complex of problems regarding the syphilis concept—for instance, its relation to tabes or to progressive paralysis. These latter problems were not tackled until pathogenesis and etiology had become sufficiently developed during the second half of the nineteenth and the early twentieth century.

If we consider purely theoretically the viewpoint held during the eighteenth and the first half of the nineteenth century, we must make the following observations.

The concept of syphilis concerns us here only insofar as it alone can indicate the relation between syphilis and the Wassermann reaction; but the concept is also defined in turn by statements that connect a certain number of other concepts. If we look at the various concepts of syphilis described here—(1) the concept of carnal scourge; (2) the empirical-therapeutic concept (mercury); (3) the experimental-pathological concepts of (a) the unitarians, (b) the dualists, (c) the adherents of the identity theory, and so on—just in terms of formal structure, independently of cultural-historical connections, it might seem at first glance to involve only an argument about a conventional definition. All these points of view are based upon observation, perhaps even upon experiments, and none can simply be declared wrong. Yet, if syphilis can be defined in various ways, the definition selected still determines some conclusions. In this respect a certain amount of latitude appears to exist. It is only after the choice has been made that the associations produced by it are seen as necessary. As is well known, this is a viewpoint held by the conventionalists. For instance, it is a matter of free choice to define syphilis or the great pox simply as the carnal scourge. But this would necessarily imply the inclusion of gonorrhea, soft chancre and so on, as well as the abandonment of a therapeutic complex, and possibly of a rational method of treatment altogether. Alternatively a definition could be constructed that is based on the usefulness of Hg. A very practical therapeutic concept would thus be obtained for what we now call the primary and secondary stages, but the tertiary stage and the metasyphilitic diseases would be excluded from this relation. Although the unitarians and others would be obliged to accept a very complicated convention, even here a description could be constructed that is congruent with their postulates.

From this formal point of view, therefore, there are some associations which are open to choice, that is, free associations, and others that are constrained. Those who recognize economy of thought[18] as the intention to choose from among the active free associations are guided by the theory of Mach.

First, the adherents of all these formal points of view pay far too little, if any, attention to the cultural-historical dependence of such an alleged epistemological choice—the alleged convention. Sixteenth-century physicians were by no means at liberty to replace the mystical-ethical concept of syphilis with one based upon natural science and pathogenesis. A stylistic bond exists between many, if not all, concepts of a period, based on their mutual influence. We can therefore speak of a thought style which determines the formulation of every concept. History shows that violent arguments can rage over the definition of concepts. This demonstrates quite independently of any utilitarian reasons just how little such conventions, which from the point of view of logic may seem equally possible, are in fact felt to be of equal value.

Second, we can find specific historical laws governing the development of ideas, that is, characteristic general phenomena concerning the history of knowledge, which become evident to anyone who examines the development of ideas. For instance, many theories pass through two periods: a classical one during which everything is in striking agreement, followed by a second period during which the exceptions begin to come to the fore. It is also evident that some ideas appear far in advance of their rationale and independently of it. Again, the interweaving of a few strands of ideas can produce special phenomena. Last, the more systematically developed, the richer both in detail and in its relations to other branches a given branch of knowledge is, the fewer will be the differences of opinion in it.

If these general cultural-historical relations as well as special ones in the history of knowledge are taken into account, conventionalism will be considerably restricted. Free rational choice or convention will be replaced by the special conditions of which we have just spoken. Nonetheless, there are always other connections which are also to be found in the content of knowledge that are not explicable in terms either of psychology (both individual and collective) or of history. For this very reason these seem to be "real," "objective," and "true" relations. We call them the passive connections in contrast with the others which we call active. In our history of syphilis the combination of all venereal diseases under the generic concept of carnal scourge was thus an *active association of the phenomena,* explained in terms of cultural history. In contrast, a restriction of the curative effect of mercury in the sentence quoted earlier asserting that "sometimes mercury does not cure the carnal scourge but makes it even worse" represents a *passive association* with respect to the act of cognition. It is of course evident that this passive association *alone* could not even be formulated if it were not for the concept of carnal scourge and that, incidentally, the very concept of carnal scourge contains passive as well as active elements.

Besides this theory about active and passive associations and their being inevitably united, the history of the development of the syphilis concept thus far shows the limited significance of any single experiment compared with the total experience consisting of experiments, observations, skills, and transformation of concepts available within a given field. Even a heroic "crucial experiment," such as that performed by Hunter, proves nothing, for its result must now be regarded as either an accident or an error. Today we know that greater experience in the field of inoculation would soon have shown Hunter the need to revise his views.

There is, however, a very important difference between experiment and experience thus construed. Whereas an experiment can be interpreted in terms of a simple question and answer, experience must be understood as a complex state of intellectual training based upon the interaction involving the knower, that which he already knows, and that which he has yet to learn. The acquisition of physical and psychological skills, the amassing of a certain number of observations and experiments, the ability to mold concepts, however, introduce all kinds of factors that cannot be regulated by formal logic. Indeed, such interactions as those mentioned prohibit any systematic treatment of the cognitive process.

There is therefore no raison d'être for any speculative epistemology, even if it be regarded as a deduction from several
examples. A great deal still remains to be investigated empirically
and discovered about the process of cognition.

Returning to our subject and specifically to the further history of
the concept of syphilis, we must mention two other ideas which
advanced it to its current form. These are the idea of syphilis as a
pathogenetic disease entity, in the extended sense of the word,
and that of the special etiological entity.

Pathogenetic ideas about syphilis, that is, opinions about the
mechanism of the pathological associations, appear in the earliest
treatises on syphilis. Almost invariably these were based on the
theory of dyscrasia, which involved noxious, foul mixtures of
humors. The whole of medicine was dominated by this theory—or
rather this empty phrase, for it permitted only about ten possibilities of combination, as if these were sufficient to cover all
diseases. It would be beyond the scope of the present work to
describe its peripeteia in detail, but one important aspect ought
to be emphasized: the idea of foul blood in syphilitics developed
from this general theory of the mixture of humors.

"Change in the blood" was a popular phrase used to explain all
generalized diseases.[19] Whereas it went progressively out of fashion
for other diseases, however, its significance only increased in the
case of syphilis.

Originally one could read phrases such as, "Especially when the
bones or muscles and nerves are nourished with melancholic blood
which, because it is infected with a noxious property, is not properly transformed into a nourishing substance, it thus happens that
secretions are greatly increased and where they accumulate are the
cause of the pain already mentioned."[20] This is an explanation of
the ache in the bones of a syphilitic. Again we read, "as during
epidemic fevers a mysterious bad quality in the air corrupts the
heart itself, the breath, and the blood."[21] Similarly, "The blood,
specifically of the syphilitic, is converted from its good state to a
bad and unnatural one."[22] Or, "When opened up here, it is clearly
seen that scabs and ulcers are beneath the surface. The cause is
really excessively hot and thick blood, infected with a poisonous

property."[23] Or, "It is agreed that it is not really different in those that suffer from the French pox, because from the very onset of this disease the blood is befouled by an infection attacking it without festering, and therefore relatively unnoticed."[24] Or, "The French pox is a condition caused by a universal infection in the mass of the blood" (Cataneus).[25] Or, "The blood, leaving its natural state, is changed" (Fallopio).[26]

Now syphilis is an extremely pleomorphic disease of many aspects. We often read in early treatises that it is a "proteiform" disease, since with its many forms, it reminds one of "Proteus or Chameleon."[27] Bloch writes that there was hardly any disease or symptom that was not attributed to syphilis.[28] In the search for a common factor and specific feature, attention was focused on befouled blood.

"Attempts to establish a diagnosis of syphilis on the basis of blood go back to the time when knowledge of the pathology of this disease was acquiring a more definite structure and the enormous complexity of the clinical aspect was becoming more and more obvious."[29]

"Early on, the infective agent was thought to be an acidic, corrosive fluid which, admixed to the blood, produces the independent form."[30] "Later, when it became increasingly accepted that syphilis depends upon modifications in the blood as well as upon other humors,"[31] the syphilitic rash was regarded as nature's attempt "to find a means to remove the pathogenic substance"[32] through the skin. "The French pox consists of boils caused by the varied corruption of humors" (Leonicenus).[33] Recovery was seen as a cleansing or sweetening of the blood. "For the limbs reject, when it arrives, the infected blood that is allocated for their nourishment, and this is expelled from the whole body by natural means through the skin acting like a handkerchief. This is the source of the first onset, a defilement of the skin, which is followed by the growth of putrid blisters as well as roughness and even further defilement of the skin" (Cataneus).[34]

About 1867 Geigel wrote: "That the blood as the general store of nutrition undergoes certain material changes during the course of syphilis and, furthermore, that these changes are not the same in all phases of syphilis we may correctly conclude from such anomalies in the pattern of nutrition as can be explained only on this basis."[35]

Reich in 1894, after listing all possible and impossible symptoms of syphilis, wrote, "All this is necessarily the outcome of the changed chemistry of the blood." "The blood of syphilitics definitely differs from that of healthy persons, as has already been confirmed indirectly by various manifestations of the disease and demonstrated by E. J. Gautier, who found diminished proportions of water and sodium chloride."[36] It was only about this time that the specific idea of syphilitic blood had begun to come into effect.

Hermann, whom we have already met as a kind of asocial outsider who made Homeric onslaughts against the "dogma of syphilitic blood," described a few contemporary experiments, designed to prove the syphilitic change in the blood. So there obviously *were* experimental attempts to transmit syphilis by means of blood.[37] "Another argument advanced in favor of syphilitic blood is the allegation that syphilis can be transmitted through cowpox vaccination."[38] Hermann also related that, at a meeting of the society of physicians in Vienna on 12 January 1872, "a young son of Aesculapius, Dr. Lostorfer, asserted that the reason why all previous blood tests had not yielded any tangible results was because faulty methods had been used. He claimed to be the discoverer, or, more correctly, the one who postulated the syphilis corpuscles, which were supposed to be present only in the blood of syphilitics and the existence of which in the blood rendered the diagnosis of constitutional syphilis accurate in every respect." But just a few days later this method was proved to be erroneous, because such syphilitic blood corpuscles "were by no means a specific symptom of syphilis." This also indicates that already at the time there was "examination of the blood of syphilitics with all available chemical and microscopical aids."[39]

Bruck reported on this subject in greater detail. "Even the numerous earlier, biological-chemical examinations of syphilitic blood had failed to yield any diagnostically usable results. After the experiments by Neumann-Konried, Reiss, Stonkovenoff-Seleneff, Liégeois, Malassez, Rille, Oppenheim, and Löwenbach, it was no longer possible to use changes in the number of blood corpuscles or in the hemoglobin and iron content for diagnostic purposes. Nagelschmidt was unable to confirm reduction in the resistance of the erythrocytes in syphilis, as was claimed by Monod, Verrotti, Sorrentino, and especially by Justus, which is said to manifest itself in a decrease of the hemoglobin content after the first mercury injection. Investigations concerning an increase in the albumin content of syphilitic blood (Ricord, Grassi, and others) and about such factors as reaction changes and freezing point determinations were also unsuccessful. The work of Detre and Sellei on the agglutinability of syphilitic and normal blood belongs to modern immunology. But even this as well as that of Nagelschmidt concerning the effect of syphilis serum on agglutination, hemolysis, and precipitation yielded no practical results."[40]

With amazing and unprecedented persistence, all possible methods were tried to confirm and to realize the traditional concept of syphilitic blood. It was with the so-called Wassermann reaction that success was at long last achieved. This discovery initiated some very important lines of research; and without much exaggeration it can be considered an epoch-making achievement.

First of all, it redefined syphilis, mainly in the secondary and tertiary stage, and especially in the area of so-called metasyphilitic diseases, such as tabes dorsalis and progressive paralysis. This was followed by a solution to the problems of hereditary and of latent syphilis. As a result of close cooperation with research in other fields it also disposed of fanciful relations with other diseases such as phthisis, rickets, and lupus.

But the Wassermann reaction also created and developed a discipline of its own: serology as a science in its own right. The original connection between serology and the Wassermann reaction still survives in popular everyday medical terminology. The Wassermann reaction is often referred to simply as the "serological test."

At the same time the etiological concept of syphilology became effective and was used to define the disease entity in the primary stage. This completes the present-day (!) definition of syphilis.

It is very difficult, if not impossible, to give an accurate historical account of a scientific discipline. Many developing strands of thought intersect and interact with one another. All of these would have to be represented, first, as continuous lines of development and, second, in every one of their mutual connections. Third, the main direction of the development, taken as an idealized average, would have to be drawn separately and at the same time. It is as if we wanted to record in writing the natural course of an excited conversation among several persons all speaking simultaneously among themselves and each clamoring to make himself heard, yet which nevertheless permitted a consensus to crystallize. The continuity in time of the line of thought already mapped out must continually be interrupted to introduce other lines. The main line of development often must be held in abeyance to explicate connections. Moreover, a great deal has to be omitted to preserve the idealized main line. Instead of a description of dynamic interactions, one is left with a more or less artificial scheme.

If I wished to trace the crystallization of the idea of a pathogenic agent from that of a mystical-symbolic spirit of the disease and a disease helminth through the idea of a disease poison and the contagium-vivum concept and on to the modern idea of bacteria as causative agents, I would have to begin again in the remote past. I would have to show how the idea of a pathogenic agent first came into contact with that of syphilis, then diverged from it for a time, made renewed contact in a new form, and eventually became permanently linked with it.

But a detailed description of this particular situation is unnecessary, if only because it resembles the development of the idea of syphilitic blood already discussed and offers few new facts to the theory of knowledge. One important difference, however, is worth mentioning. In advance of direct evidence for the existence of specific causative agents, indirect evidence was already available, since the contagious nature of the disease manifested itself in observation as well as in experiments. Analogies were found with other fields of pathology where the idea of a causative agent had already had a favorable effect during an era when bacteria were "popular." The discovery of the causative agent of syphilis is actually to be attributed mainly to bacteriologists active in other fields. Conversely, the Wassermann reaction was a direct result of syphilology and subsequently developed into a separate science—serology.

The discovery of the causative agent, *Spirochaeta pallida,* was the result of steady, systematic work by civil servants. After various unsuccessful experiments by others to discover the syphilis agent, "J. Siegel in 1904 and 1905 described structures in various contagious diseases—smallpox, foot-and-mouth disease, scarlet fever, and syphilis. He interpreted these as the still unknown agents causing such diseases and believed these must be construed as Protozoa. In view of the importance which would be attached to Siegel's findings, if confirmed, the president of the German Imperial Health Authority, Dr. Koehler, thought it advisable to have an independent criterion of assessment based upon tests to be carried out under the aegis of his own department."[41] "After a meeting under Koehler's chairmanship on 15 February 1905, Dr. Schaudinn, government board official of the Health Authority, accompanied by Dr. Neufeld, the then acting assistant, went to the director of the Royal University Clinic for Skin and Venereal Diseases, Professor Lesser, to enquire on behalf of the president whether Professor Lesser would be prepared to assist the Health Authority with pathological material in its investigations concerning the syphilis agent. Professor Lesser agreed and proposed his senior assistant, staff surgeon Hoffmann, as an additional member of the project." Already on 3 March, Schaudinn had succeeded in detecting, in the fresh tissue fluid of a syphilitic papule, "very delicate, vigorously moving spirochaetes, clearly visible only with the best optical aids." He designated these *Spirochaeta pallida,* distinguishing them from the coarser forms "quite often occurring on the buccal and the genital mucosa." Experiments of transmitting material containing spirochaetes to monkeys soon followed, yielding positive results. Nevertheless, although *Spirochaeta pallida* had already been found "by more than a hundred authors in the most diverse products of syphilis," the Health Authority, where the discovery had actually originated, was very reticent. "A report by the Health Authority dated 12 August 1905, drawn up by Prowazek and checked and signed by Schaudinn as correspondent, addressed to the Secretary of State of the Interior, ventures only that to see in *Spirochaeta pallida* the causative agent of syphilis is a not unjustified conclusion."* This team of civil servants, to which

---

*This contrasts with the account of Hoffmann, both in dating and in the sense of certitude exhibited by the group.—Eds.

the title "discoverer of the syphilis agent" should properly be awarded, carried out its work and judged its own results in the very same careful, rational, and conscientious manner in which it is now related by the team's intellectual successors.

From the production of pure cultures of *Spirochaeta pallida* and inoculation experiments with rabbits and monkeys, the idea of this as the causative agent was confirmed.

The modern concept of syphilis was thus fully established. The agents causing gonorrhea and soft chancre had been discovered earlier, so that these two diseases could be excluded from the picture of syphilis. *Spirochaeta pallida,* together with the Wassermann reaction, helped to classify tabes dorsalis and progressive paralysis definitely with syphilis. Since this spirochaete was found in the lymphatic ducts very soon after infection, even the first stage of syphilis was no longer regarded as a localized disease.

The four lines of thought intertwining to form the modern aspect of syphilis continued to develop as follows. Venereal disease, or "lues venerea," dominated and became the generic term. The connection with the sex act was translated from the mystical-ethical domain into straightforward physical terms. Not long ago another disease entity, lymphogranuloma inguinale, was differentiated from syphilis or, at least, more clearly defined. In this case Frei showed that the so-called skin test originating in the tbc theory takes the place of the Wassermann reaction. Research is currently being conducted into the causative agent. It is very likely that several more venereal disease entities are still waiting to be discovered. We still speak of so-called nonspecific ulcerations of the genitals and in many individual cases the diagnostic difficulties are great. Diagnoses such as pseudo-ulcus molle and pseudo-syphiloma are still used for want of better terms. A few tropical diseases are thought to be sexually transmitted. From the mercury idea a general chemotherapeutic theory arose which has contributed such wonderful remedies as Salvarsan among others. Although applied to many other fields, this theory is still most effective particularly with syphilis and related protozoic diseases.

The further development of the third line of thought—the idea of syphilitic blood—will presently be considered in detail.

A few very important points remain to be made concerning the idea of the causative agent. Several aspects of the disease are linked to the biological structure of *Spirochaeta pallida*. Special neurotropic and dermotropic viruses are suspected variants of *Spirochaeta pallida* and said to be related to the clinical course of the disease.[42] Attempts have been made to explain the syphilitic stages and relapses as the manifestation of a kind of generation change within the causative agent. Besides syphilis there are other important phenomena in the fields of pathology and epidemiology, as well as in bacteriology as an independent science, which even after this brief span of time already display a certain divergence between the development of the concept of any disease and that of its causative micro-organism.

A good example is the "infection inapparente" (Nicolle), which progresses without clinical illness and is very important in other diseases, such as typhus. Another, probably related, is that of the completely harmless germ carrier who harbors certain bacteria (diphtheria bacillus, meningococcus, for example) far more often than the associated disease.

The presence of a micro-organism is therefore not identical with its host's feeling ill. Consequently, the idea of the causative agent has lost the overriding importance it enjoyed during the classical period of bacteriology. Early theories, such as Pettenkofer's, are accordingly being revived. Today it can be claimed almost with impunity that the "causative agent" is but one symptom, and not even the most important, among several indicative of a disease; its presence alone is insufficient and because of the ubiquity of many microbes it occurs automatically when other conditions exist.

Theoretical bacteriology has further inherent difficulties. The biological character of *Spirochaeta pallida* is closely related or similar to that of *Spirochaeta cuniculi*, *Spirochaeta pallidula*, *Spirochaeta dentium*, and others. It can be distinguished only by means of tests on animals.[43] *Spirochaeta pallida* should therefore be defined by syphilis instead of the other way round. A botanical classification of the spirochaete species is no more successful than that of most other bacteria. To the extent that species can be defined in bacteriology, there is often no convergence between pathology and bacteriology, as shown, for instance, in the theory of vibriones.[44]

An additional factor is the extreme degree of variation in bacteria, which in some families (bacilli of the diphtheria-pseudo-diphtheria group, for example) is so great that, for the time being, classification of such species is out of the question.

Unpredictable fluctuations in virulence, such as transformations of the saprophytes into parasites and vice versa, altogether destroy the relation which initially appeared so simple between a given type of bacterium and its associated disease. Uhlenhuth and Zülzer are said to have recently succeeded in converting harmless water spirochaetes into virulent ones by passing them through guinea pigs.

It therefore cannot be claimed that syphilis is definable epistemologically solely on the basis of *Spirochaeta pallida*. The idea of the syphilis agent leads into uncertainties attending the concept of bacteriological species as such and will thus depend upon whatever future developments there may be in this field.

The development of the concept of syphilis as a specific disease is thus incomplete in principle, involved as it is in subsequent discoveries and new features of pathology, microbiology, and epidemiology.[45] In the course of time, the character of the concept has changed from the mystical, through the empirical and generally pathogenetical, to the mainly etiological. This transformation has generated a rich fund of fresh detail, and many details of the original theory were lost in the process. So we are currently learning and teaching very little, if anything at all, about the dependence of syphilis upon climate, season, or the general constitution of the patient. Yet earlier writings contain many such observations. As the concept of syphilis changed, however, new problems arose and new fields of knowledge were established, so that nothing here was really completed.

| Two | **Epistemological Conclusions from the Established History of a Concept** |

1. General Observations

The history of any scientific concept could be immaterial to those epistemologists who consider, for instance, the errors of Robert Mayer of no significance to the development of the law of conservation of energy.

Against this we would argue that there is probably no such thing as complete error or complete truth. Sooner or later a modification of the law of conservation of energy will prove necessary, and then we will perhaps be obliged to fall back upon an abandoned "error."

Furthermore, whether we like it or not, we can never sever our links with the past, complete with all its errors. It survives in accepted concepts, in the presentation of problems, in the syllabus of formal education, in everyday life, as well as in language and institutions. Concepts are not spontaneously created but are determined by their "ancestors." That which has occurred in the past is a greater cause of insecurity—rather, it only *becomes* a cause of insecurity—when our ties with it remain unconscious and unknown.

Biology taught me that a field undergoing development should be investigated always from the viewpoint of its past development.

Who today would study anatomy without embryology? In exactly
the same way epistemology without historical and comparative
investigations is no more than an empty play on words or an
epistemology of the imagination.

It is nonsense to think that the history of cognition has as little to
do with science as, for example, the history of the telephone with
telephone conversations. At least three-quarters if not the entire
content of science is conditioned by the history of ideas, psychology,
and the sociology of ideas and is thus explicable in these terms.

In the context of our special investigation, I believe that the
concept of syphilis is unattainable except through a study of its
history. It has already been demonstrated here that *Spirochaeta
pallida* alone cannot define the disease. Syphilis is not to be formulated as "the disease caused by *Spirochaeta pallida*." On the
contrary, *Spirochaeta pallida* must be designated "the micro-
organism related to syphilis." Any other definition of this microbe
is hopeless, and further, because of the question of germ carriers,
cannot serve to define the disease unambiguously.

It is also inadequate to define syphilis phenomenologically rather
than conceptually, in the manner that animals and plants might be
defined on the basis of their characteristics. For it is naive to think
that, although its historical development has been tortuous and
complicated, we can today arrive at the concept of the disease
entity "syphilis" simply and safely merely by using current techniques of observation and experiment.

This assumption is not admissible even as a thought experiment
[*Denkexperiment*]. Current research techniques, after all, are also
the result of historical development. They are the way they are because of just this particular history. Even the modern concept of disease entity, for example, is an outcome of precisely such a development and by no means the only logical possibility. As history shows,
it is feasible to introduce completely different classifications of diseases. Furthermore, it is possible to dispense with the concept of a
disease entity altogether, and to speak only of various symptoms
and states, of various patients and incidences. This latter point of
view is by no means impracticable because, after all, the various
forms and stages as well as the various patients and constitutions
must always be treated differently. It is evident that the formation

of the concept "disease entity" involves synthesis as well as analysis, and that the current concept does not constitute the logically or essentially only possible solution.

In this context it is not possible to regard things simply as given. Experience gained over several years of working in the venereal disease section of a large city hospital convinced me that it would never occur even to a modern research worker, equipped with a complete intellectual and material armory, to isolate all these multifarious aspects and sequelae of the disease from the totality of the cases he deals with or to segregate them from complications and lump them together. Only through organized cooperative research, supported by popular knowledge and continuing over several generations, might a unified picture emerge, for the development of the disease phenomena requires decades.

Here, however, training, technical resources and the very nature of collaboration would repeatedly lead research workers back to the historical development of knowledge, since the bonds of history can never be cut.

For epistemology, it might be objected, it is not important to investigate how a connection was discovered, but only to legitimize it scientifically, prove it objectively, and construct it logically. But this could be countered as follows.

Legitimization is certainly very important in science generally and, within reasonable bounds and precision, to our case as well; otherwise syphilology would not be a branch of science. But I do not agree with the view that the sole or even most important task of epistemology consists in this kind of examination of the consistency of concepts and their interconnections within a system.

Whatever is known has always seemed systematic, proven, applicable, and evident to the knower. Every alien system of knowledge has likewise seemed contradictory, unproven, inapplicable, fanciful, or mystical. May not the time have come to assume a less egocentric, more general point of view and to speak of comparative epistemology? A rule of thought that allows one to make use of more details and more compulsory connections, as the history of science teaches us, deserves to be emphasized. I believe that the principles used in the present study render many a neglected relation both visible and suitable for investigation.

The concept of syphilis must be investigated like any other case in the history of ideas, as being a result of the development and confluence of several lines of collective thought.

It is not possible to legitimize the "existence" of syphilis in any other than a historical way. To avoid unnecessary and traditional mysticism it is thus appropriate to use the term "existence" restrictively as only a thinking aid and convenient shortcut.[1] But it would be a gross mistake merely to assert that the syphilis concept could not be attained without the consideration of particular historical connections. We still have to examine possible laws behind these connections and discover operative socio-cogitative forces.

2. Proto-ideas as Guidelines for the Development of any Finding

Many very solidly established scientific facts are undeniably linked, in their development, to prescientific, somewhat hazy, related proto-ideas or pre-ideas, even though such links cannot be substantiated.

We have described a hazy idea of syphilitic changes in the blood and shown that this idea existed centuries before scientific proof was available. Emerging from a chaotic mixture of ideas, it developed over many epochs, becoming more and more substantial and precise. Evidence for it was adduced from various points of view, and a dogma concerning syphilitic blood gradually consolidated. Several research workers, such as Gautier, succumbed to public opinion and claimed to have found proofs which actually were impossible to establish. The entire repertoire of research available at the time was used to an unprecedented degree until the goal was reached. The idea of syphilitic blood thus became scientifically embodied in the Wassermann reaction and subsequently in more simplified reactions. But the proto-idea has survived among the common people, who still refer to syphilitics as having impure blood.

Seen from this point of view, the Wassermann reaction in its relation to syphilis constitutes the modern, scientific expression of an earlier pre-idea which contributed to the concept of syphilis. Other branches of science also have pre-ideas. The pre-idea of atomic theory stems from Greek antiquity, specifically as taught by Democritus in his original "atomistics." Historians of science such as Paul Kirchberger[2] and Friedrich Lange agree that "modern atomic theory developed step by step from the atomistics of Democritus."[3] It is a constant source of amazement to see just how many features of modern atomic theory were preformed in the theses of these ancient precursors, such as the combination and separation of atoms, mutual gravitational motions and their effects, as well as pressure and collision phenomena.

Theories of the elements and of chemical composition, the law of conservation of matter, the principle of a spherical earth as well as the heliocentric system each underwent a historical development from somewhat hazy proto-ideas. These existed long before any scientific proofs were available and were supported in different ways throughout the intervening period until they received a modern expression.

Very clear ideas about tiny invisible and living agents as causative of diseases were expressed long before the advent of the modern theory of infection and even before the invention of the microscope. A statement from Varro, "Minute animals that cannot be seen by the eye enter the body from the air through the mouth and also through the nose and cause severe diseases," might have come from a popular edition of Flügge's theory of droplet infection.*

I admit that sometimes a proto-idea could be found for a scientific discovery only through casuistry. We would look in vain for such a proto-idea in the case of isomerism or in the gram differentiation of bacteria. Nor must every ancient idea have a historical relation with a later discovery which it happens to resemble. The Aschheim-Zondek test for pregnancy, for example, is probably unrelated to the medieval idea of the possibility of diagnosing virginity or pregnancy from urine. In spite of prolonged investigation, some ideas remained devoid of scientific proof and were eventually discarded. After just such a search for the "absolute," today there is not even a scientific term to denote it clearly.

Can epistemology blandly ignore the fact that many scientific positions steadily developed from proto-ideas which at the time

*Fleck is referring to Flügge's *Die Mikroorganismen* (see Bibliography).—Eds.

were not based upon the type of proof considered valid today? This question should be reflected upon and investigated. But if we may borrow a hypothesis from the prehistory of paleontology, a proto-idea must not be construed as a "freak of nature." Proto-ideas must be regarded as developmental rudiments of modern theories and as originating from a socio-cogitative foundation.

It might be claimed that, whereas a large number of somewhat hazy ideas have emerged throughout history, it is science that has accepted the "right" ones and rejected the "wrong" ones. But this objection is untenable, since it cannot explain why there are so many possible "correct" representations of unknown objects. Implicit in such a view is the claim that the categories of truth and falsehood may be applied to these proto-ideas. But this suggestion is altogether erroneous. Was "befouled blood," "corrupted or melancholic blood," or "excessively hot and thick" the correct idea for syphilis? "Befouled" is not a precise scientific term. Because it is vague and ambiguous, we cannot decide whether it is suitable for syphilis or not. It is not a systematic term as required today, although it was clearly useful as a starting point in the development of a concept. Even the most suitable of the earlier descriptions—"change in the blood"—can no longer be checked for accuracy. The characteristic "change" is too vague, and a "change in the blood" could correspond in one sense or another to almost any condition or disease. Moreover, "syphilis" means something entirely different today from what it meant formerly. The value of such a pre-idea resides neither in its inner logic nor in its "objective" content as such, but solely in the heuristic significance which it has in the natural tendency of development. And there is no doubt that a fact develops step by step from this hazy proto-idea, which is neither right nor wrong.

Concerning other proto-ideas, such as the Greek pre-idea of the atom or that of the elements, we are also unable to decide whether they are right or wrong if they are taken out of their chronological context, because they correspond to a different thought collective and a different thought style. Although such ideas may not conform to modern scientific thought, their originators certainly considered them to be correct.

Any absolute criterion of judgment as to suitability is as invalid for fossilized theories as a chronologically independent criterion would be for adaptability of some paleontological species. The brontosaurus was as suitably organized for its environment as the modern lizard is for its own. If considered outside its proper environmental context, however, it could not be called either "adapted" or "unadapted."

The development of thought proceeds so much more rapidly than the pace appropriate to paleontology that we continuously witness the occurrence of "mutations" in thought style. The transformation in physics and in its thought style brought about by relativity theory represents such a mutation, as does the adjustment in bacteriology resulting from the theories of variability and cyclogeny. Suddenly we no longer see clearly what is species and what is individual, or how broadly the concept of life cycle is to be taken. What just a few years ago was regarded as a natural event appears to us today as a complex of artefacts. Soon we shall no longer be able to say even whether Koch's theory is correct or not, because new concepts incongruent with Koch's will arise from the present confusion.

Another comparison taken from the area of word origin, as recently conjectured by some psychologists, may perhaps better explain the importance of pre-ideas. "Words were originally not phonetic nexuses arbitrarily assigned to certain objects, such as the word UFA* denoting a German film studio or 'L' denoting self-induction. They actually indicate a transference of experience and objects to a material that can easily be molded and is always available. Linguistic reproduction was therefore originally not a precise assignment according to logic but imagery in the dynamic sense of geometry. The meaning would be immediately implicit in ideophones created in this way."[4] The actuality of pre-ideas probably permits the assumption of a similar relationship. Mental reproduction would be originally not a clear-cut assignment according to logic, but rather a transference of experience to a material that could easily be molded and would always be available. The connection between reproduction and experience would not be like the conventional relation between a symbol and what it symbolizes, but would reside in a psychological correspondence

*Universum Film AG, which is a studio, like MGM.—Eds.

between the two. Evidence for this would be directly contained in the products of thought [*Denkgebilden*] created in this way.

Words, then, were not originally names for objects. And cognition, at least initially, does not depend upon mental reconstruction and preformation of phenomena or, as Mach taught,[5] upon the adaptation of thoughts to some arbitrary external facts as revealed to an average person.

Words and ideas are originally phonetic and mental equivalents of the experiences coinciding with them. This explains the magical meaning of words and the dogmatic, reverential meaning of statements.

Such proto-ideas are at first always too broad and insufficiently specialized. According to Hornbostel, ideas—just as word meanings—have a development that proceeds "not through abstraction from the particular to the general, but through differentiation or specialization from the general to the particular."

3. The Tenacity of Systems of Opinion and the Harmony of Illusions; Viewpoints as Autonomous, Style-Permeated Structures [Gebilde]

Once a structurally complete and closed system of opinions consisting of many details and relations has been formed, it offers enduring resistance to anything that contradicts it.

A striking example of this tendency is given by our history of the concept of "carnal scourge" in its prolonged endurance against every new notion. What we are faced with here is not so much simple passivity or mistrust of new ideas as an active approach which can be divided into several stages. (1) A contradiction to the system appears unthinkable. (2) What does not fit into the system remains unseen; (3) alternatively, if it is noticed, either it is kept secret, or (4) laborious efforts are made to explain an exception in terms that do not contradict the system. (5) Despite the legitimate claims of contradictory views, one tends to see, describe, or even illustrate those circumstances which corroborate current views and thereby give them substance.

In the history of scientific knowledge, no formal relation of logic exists between conceptions and evidence. Evidence conforms to conceptions just as often as conceptions conform to evidence. After all, conceptions are not logical systems, no matter how much they aspire to that status. They are stylized units which either develop or atrophy just as they are or merge with their proofs into others. Analogously to social structures, every age has its own dominant conceptions as well as remnants of past ones and rudiments of those of the future. It is one of the most important tasks in comparative epistemology to find out how conceptions and hazy ideas pass from one thought style to another, how they emerge as spontaneously generated pre-ideas, and how they are preserved as enduring, rigid structures [*Gebilde*] owing to a kind of harmony of illusions. It is only by such a comparison and investigation of the relevant interrelations that we can begin to understand our own era.

To clarify the point a few examples might be mentioned showing the various degrees of tenacity of viewpoints.

1. When a conception permeates a thought collective strongly enough, so that it penetrates as far as everyday life and idiom and has become a viewpoint in the literal sense of the word, any contradiction appears unthinkable and unimaginable. People argued against Columbus: "Could anyone be mad enough to believe that there are antipodes; people standing with their feet opposite our own, who walk with their legs sticking up and their heads hanging down? Is there really a region on earth where things are upside down, where trees grow downward, and where it rains, hails, and snows upward? The delusion that the earth is round is the cause of this foolish fable."

Today we know that the real cause of difficulty here was the absolute meaning of the concepts "up" and "down"—a problem that dissolves under a relativistic formulation. The same difficulty arises even today if such concepts as existence, reality, and truth are used in an absolute sense. To Kant, an unknowable substratum as "things in themselves" was indispensable for sensory appearances: "otherwise we should be landed in the absurd conclusion that there can be appearance without anything that appears."[6] Wundt concurs by asking, "What can one do with properties and states which are not properties and states of something?"[7]

2. Every comprehensive theory passes first through a classical

stage, when only those facts are recognized which conform to it exactly, and then through a stage with complications, when the exceptions begin to come forward. The great theoretician Paul Ehrlich knew this only too well: "Unfortunately, this differs in no way from all other scientific problems, since it just becomes more and more complicated."[8] In the end there are often more exceptions than normal instances.

Such a relation exists between classical chemistry and the chemistry of colloids. Colloidal reactions vastly predominate in nature over classical chemical reactions. Nevertheless, like the colloidal reactions, those events which occur with greater frequency have often had to wait longer for scientific discovery. Many aspects of tanning, dyeing, and the production of adhesives, rubber, and explosives do not correspond to the laws of classical chemistry. Furthermore, special laws must be assumed to explain how agricultural soil can retain nutrient salts, which according to classical chemical and physical laws should be washed away freely by the groundwater. All these many "exceptions" went unrecognized for a long time.

Another instructive example is the fate of observations made in 1908 by Bjerrum and Hantzsch. These seemed to contradict the classical theory of electrolytic dissociation and thus had to wait about ten years until they were repeated by other workers. Proper recognition of these observations obtained only after publication of the work of Laue and Bragg. The simple fact went unnoticed that the color of an ionic salt solution can, during dilution, be so modified that the degree of dissociation appears to remain unchanged. Nor was any attention paid to the fact that the addition of $CaCl_2$ to salt solutions displaces the normal reaction of the mixture in the acid direction.

Take an example from everyday life. At a time when sexuality was equivalent to uncleanness and naiveté to purity, naive children were thought to be asexual. How amusing it is that sexuality could not be recognized in them! Everyone has the experience of having himself been a child and now lives not entirely isolated from children. Yet it took psychoanalysts to discover children's sexuality.

We see the same thing happening in the classical theory of infectious diseases. Every infectious disease was supposed to be caused by very small living "agents." Nobody could see that these "agents" were also present in healthy persons, because the phenomenon of the germ carrier was not discovered until much later. The variability of micro-organisms was a second shock. At the time when Koch's theory of specificity held complete sway, any variability was unthinkable.[9] It was only some time later that relevant observations became more frequent. A third shock will be administered to the classical theory of infection by the theory of the filterable virus. It will be shown that invasion by a causative agent, which is the classic cause of infection, is actually an exceptional way to produce an infection.

This particular example well illustrates the important role that the tenacity of closed systems of opinion plays in the operation of cognition [*Erkenntnisphysiologie*]. Cognition proceeds in this and in no other way. Only a classical theory with associated ideas which are plausible (rooted in the given era), closed (limited), and suitable for publication (stylistically relevant) has the strength to advance. Loeffler's bacilli, for instance, would never have been isolated had they first been found in healthy persons. Without a function in an era preoccupied with "causes," this finding would never have attracted the necessary attention nor stimulated the necessary research effort.

Discovery is thus inextricably interwoven with what is known as error. To recognize a certain relation, many another relation must be misunderstood, denied, or overlooked.

The operation of cognition [*Erkenntnisphysiologie*] is analogous to the physiology of movement. To move a limb, an entire so-called myostatic system must be immobilized to provide a basis of fixation. Every movement consists of two active processes; namely, motion and inhibition. The corresponding features in the operation of cognition are purposive, directed determination and cooperative abstraction, which complement one another.

3. We have mentioned concealment of an "exception" among the stages of tenacity in opinion systems. One of many exceptions was the orbital motion of Mercury as related to Newton's laws. Experts in the field were aware of it, but it was concealed from the public because it contradicted prevailing views. It is mentioned only now because it became useful in the context of relativity theory.

4. The very persistence with which observations contradicting a

view are "explained" and smoothed over by conciliators is most
instructive. Such effort demonstrates that the aim is logical conformity within a system at any cost, and shows how logic can be
interpreted in practice. Every theory aspires to being a logical
system but often merely begs the question.

The following passage from Paracelsus is so much to the point
here that to quote it here will spare the reader many examples.[10]

Man, who alone walks in the visible light of nature, is unable to believe that a man could be possessed by the devil and harbor him in
such a way that one must think: This man is *not* a man, but a devil.
This possibility arouses revulsion and resentment in any living person. Must it not be a miracle of God that a man living in this world
could appear to have a devil?[11] Man is supposed to be made in the
image of God and not of the devil, who is as different from man as
stone from wood. Aside from the fact that man is made in the
image of God, he has also been redeemed from the devil by the Son
of God. How incredible therefore that nevertheless he can be
thrown into such frightful captivity without protection!

Two articles of faith confront each other here, namely, that man
could be possessed by the devil, and that yet he was freed from the
devil. Neither of these articles may be doubted, but something
must be done to save the relevant logic. A miracle of God is
invoked to bring them into accord. This saves the logic of the matter
and no one need any longer harbor "revulsion and resentment."

No matter how illogical this may appear to us, the whole thing is
true to style. Let us try to imagine ourselves in the world of
Paracelsus, where every object and event is a symbol, and every
symbol and metaphor also has objective value. It is a world full of
hidden meanings, spirits, and mysterious powers, full of defiance
and awe as well as love and hate. How else could one live in such an
impulsive, unsafe, and hazardous reality than to believe in mira-
cles? The miraculous becomes the fundamental principle and most
immediate experience within Paracelsian reality and permeates
every aspect of his science. It anticipates every consideration, and
springs forth from every consideration.

A closed, stylized system of this kind is not immediately re-
ceptive to new ideas. These would be reinterpreted to make them
conform.

5. The liveliest stage of tenacity in systems of opinion is creative fiction, constituting, as it were, the magical realization of ideas and the interpretation that individual expectations in science are actually fulfilled.

Almost any theory can be quoted as an example here, because all contain some element of wishful thinking by their scientific proponents. But concrete and detailed examples are more useful for illustrating the extent of such wishful thinking than for merely establishing its existence.

In an age when marveling at nature was sufficient to be regarded as knowledge, and before man had learned to utilize his admiration in a practical way to stimulate proper investigation, the purposefulness of things in nature, both living and inanimate, was wondered at and enormously overvalued. Marvelous instincts aroused particular fascination. In an essay "The Nests of Animals" published in 1866, Wood tells the following story.[12] "Maraldi was struck by the great regularity exhibited by bees' honeycombs. He measured the angles of the rhombohedral dividing walls and found them to be 109° 28' and 70° 32'. Convinced that these particular angles must somehow be related to the economy of the cells, Réaumur asked the mathematician König to calculate what shape a hexagonal vessel bordered by three rhombi would have if it enclosed the maximum volume with minimum surface. Réaumur received the answer that the rhombic angles would have to be 109° 26' and 70° 34', constituting a difference of only two minutes of arc. Not satisfied with this lack of agreement, Maclaurin repeated Maraldi's measurements and confirmed them. But when he repeated the calculation he spotted an error in the table of logarithms used by König. It was not the bee but the mathematician that had made the mistake. The bees even helped to discover the error." Mach also comments on this case: "Those who know how to measure crystals, and have seen a honeycomb with its rather rough and nonreflecting surfaces, will doubt that an accuracy of two minutes can be achieved in its measurement. The story must therefore be considered only a doughty legend of mathematics.... It must also be said that the mathematical problem had not been fully presented, so there is no way to judge to what extent the bees have actually solved it."

Those who find this story, written as it is in a quite scientific style,[13] insufficiently convincing to demonstrate the occurrence of self-fulfilling scientific expectation* [*Wunschtraumerfüllung*], may prefer to look at even "more objective fiction" in the form of pictorial representations.

In an Amsterdam transcription by N. Fontanus[14] of Vesalius's *Epitome*, the uterus is illustrated on page 33, with the following legend on page 32. "Question: How does the seed enter the woman during ejaculation if the womb is so tightly closed that not even a needle can enter through it, *according to Hippocrates, book 5, aphorisms 51 and 54?* Answer: Through a branch leading from the ejaculatory duct entering the cervix of the uterus, *as this illustration shows.*"

The idea of a fundamental analogy existing between male and female genitals, as held in antiquity, is exhibited most effectively here, and illustrated as if it really occurred in nature. Anatomists will notice immediately that the proportions of the organs, as well as the corresponding positioning, have been restyled to conform to this theory.[15] Truth and fiction or, perhaps better, relationships that have been retained within science and others that have disappeared from this structure appear here visibly side by side. The duct labeled *S*, "through which the woman becomes impregnated by the seed ejaculated at the time of intercourse," is typical, and it is indispensable to this theory of analogy. Although unknown in modern anatomy, it is pictured in early anatomical descriptions in a style appropriate to that theory—right amidst other excellent data of observation.

When I selected this illustration for the present work, I was tempted to add a "correct" and "faithful" one for comparison. Leafing through modern anatomical atlases and gynecological textbooks, I found many good illustrations but not a single natural one. All had been touched up in appearance, and were schematically, almost symbolically, true to theory but not to nature. I found one particular photograph in a textbook on dissecting techniques. This, too, was tailored to theory with orientation lines and indicating arrows added to make it graphically suitable for use in teach-

---

*This free, anachronistic translation of *Wunschtraumerfüllung* echoes the title of a paper by Robert K. Merton, "The Self-Fulfilling Prophecy," *Antioch Review*, summer 1948 (reprinted as chap. 13 of Merton's *Social Theory and Social Structure*, New York, 1968).—Eds.

From Andreas Vesalius' books on
the structure of the human body.
After Fontanus 1642.

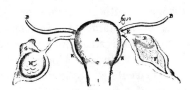

From Thomas Bartholin's
Anatomy, 1673.

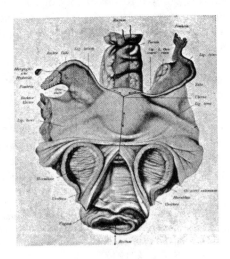

From Coelestin Nauwerck,
Technique of Dissection,
1912.

Figure 1

ing. I thus once again convinced myself that it is not possible to
carry out such a comparison. It is only theories, not illustrations,
that can be compared. It is true that modern doctrine is supported
by much more sophisticated techniques of investigation, much
broader experience, and more thorough theory. The naive analogy
between the organs of both sexes has disappeared, and far more
details are available. But the path from dissection to formulated
theory is extremely complicated, indirect, and culturally conditioned. The more clearly we visualize it, the more we will be
confronted with connections in the history of ideas and psychology
leading us to their originators. In science, just as in art and in life,
only that which is true to culture is true to nature.

Any attempt to legitimize a particular approach as the correct
one is at best of limited value, since it is inextricably bound to a
thought collective. Neither the style characteristic of opinion nor
the technical skills required for any scientific investigation can be
formulated in terms of logic. This sort of legitimization is therefore
possible only where it is actually no longer required, namely among
persons whose intellectual constitution is thought-stylized in common and, more particularly, who share approximately the same
educational background.

Berengar, for instance, discusses somewhere the old argument
about the origin of the veins.[16] According to Aristotle the veins
originate in the heart, according to Galen in the liver. "I say that
the veins originate neither in the heart nor in the liver except in a
figurative and metaphorical sense, yet metaphorically they originate more in the liver than in the heart, and thus in this respect I
agree more with the physicians than with Aristotle." Here it is
obvious that any logical discussion is bound to fail. We do not
recognize such "metaphorical and figurative" origins for the veins,
but only a morphological, phylogenetic, and embryological "origin" of the blood vessels. To us the human body does not represent
such a collection of metaphors and symbols, although we are unable
to provide a logical reason why we have changed the style of
approach.

Simple lack of "direct contact with nature" during experimental
dissection cannot explain the frequency of the phrase "which becomes visible during autopsy" often accompanying what to us seem

the most absurd assertions. Such contact was generally very tenuous. It was much less the autopsy itself than the ancient views that were consulted, yet this only served to reinforce the ancient thought style. Stereotyped opinions handed down a thousand times meant more to and were considered safer by those authors than any autopsy as such, which was only a "gruesome duty."[17]

This period was associated with a specifically symbolic "anatomia imaginabilis" and was followed by another period when a purely morphological approach to anatomy was attempted. The latter, however, could not be pursued without phylogenetic, ontogenetic, and comparative symbols.[18] Then came physiological anatomists, using physiological symbols and evolving chemical organs, the endocrine system, the reticulo-endothelial system: structures to which no sharply outlined morphological organs correspond. During each period with its own characteristic style the concepts used were absolutely clear, since clarity is based upon reducibility to other stylized concepts. Despite this clarity, direct communication between the adherents of different thought styles is impossible.[19] How is the ancient anatomical term "bosom" [Schoss],* for instance, to be translated into a modern one? Where is this mystical organ to be positioned?

An example of a nineteenth-century scientific illustration might be added, very similar to the one from the seventeenth century. When Haeckel, the romantic, high-spirited champion of truth, wanted to demonstrate his ideas about descent, he did not shrink from occasionally using the same blocks for the illustration of different objects such as animal and human embryos which should look alike according to his theory. His *History of Natural Creation* abounds with biased illustrations appropriate for his theory. Compare, for instance, the intelligent faces of the old chimpanzee and the old gorilla on figure 13 with the exaggeratedly gruesome ones of the Australian aborigine and the Papuan on figure 14. I should like to mention in conclusion a particularly gross example of the tendency to save a point of view. "The seemingly best support for the heredity of acquired characteristics comes from the experiments of

---

*The German word *Schoss* means "lap" physically but "bosom" metaphorically or mystically, as to be "safe in the bosom of Abraham."—Eds.

Kammerer. Using the effect of dampness, yellow background, and other, general factors he had altered specimens of spotted salamanders, *Salamandra maculosa,* to appear as striped ones. He excised the ovaries of these artificially striped salamanders, transplanting in their place ovaries of spotted specimens. When he mated these animals with normally spotted salamanders, the latter produced offspring with spots arranged in rows. It appeared here that the artificially altered somatic cells had influenced the egg cells of the mate." These results caused lively discussion until suddenly "Kammerer's experiments were proved to have been fraudulent (1926), a discovery which led to his suicide."[20]

To the objection that such examples, particularly the last one, are not representative of the normal function of cognition, I would admit that many of these self-fulfilling expectations could be viewed in this way. But as a physician I know that we cannot distinguish sharply between normality and abnormality. The abnormal is often only an enhancement of the normal. It is also known that both normality and abnormality often have identical social effects. Although the philosophy of Nietzsche has, for instance, a psychopathological motif, it generates social effects no different from those produced by a normally conditioned outlook on life. At any rate, once a statement is published it constitutes part of the social forces which form concepts and create habits of thought. Together with all other statements it determines "what cannot be thought in any other way." Even if a particular statement is contested, we grow up with its uncertainty which, circulating in society, reinforces its social effect.[21] It becomes a self-evident reality which, in turn, conditions our further acts of cognition. There emerges a closed, harmonious system within which the logical origin of individual elements can no longer be traced.

Every pronouncement leaves behind either the solution or the problem, if only the problem of the problem's own rationality. The formulation of a problem already contains half its solution. Any future examination must return along existing thought tracks. The future will never be completely isolated from the past, whether normal or abnormal, except when a break with it occurs as the result of the rules characteristic of the thought structure in question.

The tenancity of systems of opinion shows us that, to some extent, they must be regarded as units, as independent, style-permeated structures. They are not mere aggregates of partial propositions but as harmonious holistic units exhibit those particular stylistic properties which determine and condition every single function of cognition.

The self-contained nature of the system as well as the interaction between what is already known, what remains to be learned, and those who are to apprehend it, go to ensure harmony within the system. But at the same time they also preserve the harmony of illusions, which is quite secure within the confines of a given thought style.

4. Introduction to Thought Collectives

In comparative epistemology, cognition must not be construed as only a dual relationship between the knowing subject and the object to be known. The existing fund of knowledge must be a third partner in this relation as a basic factor of all new knowledge. It would otherwise remain beyond our understanding how a closed and style-permeated system of opinions could arise, and why we find, in the past, rudiments of current knowledge which at the time could not be legitimized by any "objective" reasons and which remained only pre-ideas.

Such historical and stylized relations within knowledge show that an interaction exists between that which is known and the act of cognition. What is already known influences the particular method of cognition; and cognition, in turn, enlarges, renews, and gives fresh meaning to what is already known.

Cognition is therefore not an individual process of any theoretical "particular consciousness." Rather it is the result of a social activity, since the existing stock of knowledge exceeds the range available to any one individual.

The statement, "Someone recognizes something," whether it be a relation, a fact, or an object, is therefore incomplete. It is no more meaningful as it stands than the statements, "This book is larger," or "Town A is situated to the left of town B." Something is still missing, namely the addition, "than that book," to the second statement, and either, "to someone standing on the road between towns A and B while facing north," or "to someone walking on the road from town C to town B," to the third statement. The relative terms "larger" and "left" acquire a definite meaning only in conjunction with their appropriate components.

Analogously, the statement, "Someone recognizes something," demands some such supplement as, "on the basis of a certain fund of knowledge," or, better, "as a member of a certain cultural environment," and, best, "in a particular thought style, in a particular thought collective."

If we define "thought collective" as *a community of persons mutually exchanging ideas or maintaining intellectual interaction, we will find by implication that it also provides the special "carrier" for the historical development of any field of thought, as well as for the given stock of knowledge and level of culture. This we have designated thought style.* The thought collective thus supplies the missing component.

The statement, "Schaudinn discerned *Spirochaeta pallida* as the causative agent of syphilis," is equivocal as it stands, because "syphilis as such" does not exist. There was only the then-current concept available on the basis of which Schaudinn's contribution occurred, an event that only developed this concept further. Torn from this context, "syphilis" has no specific meaning, and "discerned" by itself is no more explicit than "larger" and "left" in the examples above.

Siegel also recognized, in his own way, protozoa-like structures as the causative agent of syphilis. If his findings had had the appropriate influence and received a proper measure of publicity throughout the thought collective, the concept of syphilis would be different today. Some syphilis cases according to present-day nomenclature would then perhaps be regarded as related to variola and other diseases caused by inclusion bodies. Some other cases would be considered indicative of a constitutional disease in the strict sense of the term. Following the train of thought characterized by the "carnal scourge" idea, still another, completely different set of concepts concerning infectious disease and disease entities would have arisen. Ultimately we would still have reached a harmonious system of knowledge even along this line, but it would differ radically from the current one.

Although such a possibility could be envisioned logically and "objectively," it can never be construed as a historical possibility. By Siegel's time, the syphilis concept had already become too rigid for such a sweeping change. A century earlier, when the concept was still sufficiently adaptable, the intellectual and experimental requirements did not yet exist for Siegel's finding. We need have no scruples about declaring that the finding of Schaudinn was correct and that of Siegel incorrect. Schaudinn's was uniquely linked with a thought collective (usually there is only one such possible link), whereas Siegel's lacked such a connection. The former served as the junction for lines of development of several ideas within this collective, but the latter did not. The meaning and the truth value of Schaudinn's finding is therefore a function of the community of those who, maintaining intellectual interaction on the basis of a shared intellectual past, made his achievement possible and accepted it.

Correctly formulated, the statement describing Schaudinn's discovery would read as follows: "Conforming to then-current views about syphilis and causative agents, Schaudinn suggested that *Spirochaeta pallida* should be recognized as the causative agent of syphilis. The significance of *Spirochaeta pallida* was duly accepted, and this idea was used for the further development of syphilology." Does not every reputable textbook in bacteriology describe the circumstances in this manner?

Cognition therefore means, primarily, to ascertain those results which must follow, given certain preconditions. The preconditions correspond to active linkages and constitute that portion of cognition belonging to the collective. The constrained results correspond to passive linkages and constitute that which is experienced as objective reality. The act of ascertaining is the contribution of the individual.

The three factors involved in cognition—the individual, the collective, and objective reality (that which is to be known)—do not signify metaphysical entities; they too can be investigated, for they have further relations with respect to one another.

These further relations consist in the facts that, on the one hand, the collective is composed of individuals and that, on the other, objective reality can be resolved into historical sequences of ideas belonging to the collective. It is therefore possible from the viewpoint of comparative epistemology to eliminate one or perhaps even two factors.

Although the thought collective consists of individuals, it is not simply the aggregate sum of them. The individual within the collective is never, or hardly ever, conscious of the prevailing thought style, which almost always exerts an absolutely compulsive force upon his thinking and with which it is not possible to be at variance.

strong

The presence of thought style makes it necessary and, indeed, indispensable to construct the concept "thought collective." Whoever might nevertheless prefer to eliminate the thought collective must introduce value judgments or dogmatic faith into the theory of knowledge. In so doing he would only succeed in creating a particular dogmatic type of epistemology out of the general comparative type.

The important role of the collective effort in any scientific work is clearly shown by the history of syphilology as described in chapter 1. Every theme in the sequence of ideas originates from notions belonging to the collective. Disease as a punishment for fornication is the collective notion of a society that is religious. Disease caused by the influence of the stars is a view characteristic of the astrological fraternity. Speculations of medical practitioners about therapy with metals spawned the mercury idea. The blood idea was derived by medical theoreticians from the vox populi, "Blood is a humor with distinctive virtues."* The idea of the causative agent can be traced through the modern etiological stage as far back as the collective notion of a disease demon.

Not only the principal ideas but also all the formative stages of the syphilis concept, however, are the result of collective, not individual effort. Although we spoke of Schaudinn as the discoverer, he really no more than personified the excellent team of health officials whose work, described in the previous chapter, cannot easily be dissected for individual attribution. As we shall presently show, even the origin of the Wassermann reaction is due to a kind of experience of the collective, which actually militated

---

*Cf. Goethe's *Faust,* line 1740: "Blut ist ein ganz besondrer Saft."—Eds.

against Wassermann's views. Wassermann, like Schaudinn, is rather a standard-bearer in discovery than its sole agent.

When we look at the formal aspect of scientific activities, we cannot fail to recognize their social structure. We see organized effort of the collective involving a division of labor, cooperation, preparatory work, technical assistance, mutual exchange of ideas, and controversy. Many publications bear the names of collaborating authors. Scientific papers almost invariably indicate both the establishment and its director by name. There are groups and a hierarchy within the scientific community: followers and antagonists. societies and congresses, periodicals, and arrangements for exchange. A well-organized collective harbors a quantity of knowledge far exceeding the capacity of any one individual.

The same pertains also to the humanities, although here the organization is less developed. Any kind of learning is connected with some tradition and society, and words and customs already suffice to form a collective bond.

Cognition is the most socially-conditioned activity of man, and knowledge is the paramount social creation [*Gebilde*]. The very structure of language presents a compelling philosophy characteristic of that community, and even a single word can represent a complex theory. To whom do these philosophies and theories belong?

Thoughts pass from one individual to another, each time a little transformed, for each individual can attach to them somewhat different associations. Strictly speaking, the receiver never understands the thought exactly in the way that the transmitter intended it to be understood. After a series of such encounters, practically nothing is left of the original content. Whose thought is it that continues to circulate? It is one that obviously belongs not to any single individual but to the collective. Whether an individual construes it as truth or error, understands it correctly or not, a set of findings meanders throughout the community, becoming polished, transformed, reinforced, or attenuated, while influencing other findings, concept formation, opinions, and habits of thought. After making several rounds within the community, a finding often returns considerably changed to its originator, who reconsiders it himself in quite a different light. He either does not recognize it as his own or believes, and this happens quite often, to have originally seen it in its present form.* The history of the Wassermann reaction will afford us the opportunity to describe such meanderings in the particular case of a completely "empirical" finding.

This social character inherent in the very nature of scientific activity is not without its substantive consequences. Words which formerly were simple terms become slogans; sentences which once were simple statements become calls to battle. This completely alters their socio-cogitative value. They no longer influence the mind through their logical meaning—indeed, they often act against it—but rather they acquire a magical power and exert a mental influence simply by being used. As an example, one might consider the effect of terms such as "materialism" or "atheism," which in some countries at once discredit their proponents but in others function as essential passwords for acceptability. This magical power of slogans, with "vitalism" in biology, "specificity" in immunology, and "bacterial transformation" in bacteriology, clearly extends to the very depth of specialist research. Whenever such a term is found in a scientific text, it is not examined logically, but immediately makes either enemies or friends.

New themes such as propaganda, imitation, authority, rivalry, solidarity, enmity, and friendship begin to appear—themes which could not have been produced by the isolated thought of any individual. Every such motif acquires epistemological importance, because the entire fund of knowledge as well as the intellectual interaction within the collective take part in every single act of cognition, which is indeed fundamentally impossible without them. Every epistemological theory is trivial that does not take this sociological dependence of all cognition into account in a fundamental and detailed manner. But those who consider social dependence a necessary evil and an unfortunate human inadequacy which ought to be overcome fail to realize that without social conditioning no cognition is even possible. Indeed, the very word "cognition" acquires meaning only in connection with a thought collective.

A kind of superstitious fear prevents us from attributing that

---

*Cf. the discussion of "the retroactive effect" in R. K. Merton, *Social Theory and Social Structure* (1968), pp. 16, 17, 37.—Eds.

which is the most intimate part of human personality, namely the thought process, also to a collective.[22] A thought collective exists wherever two or more people are actually exchanging thoughts. He is a poor observer who does not notice that a stimulating conversation between two persons soon creates a condition in which each utters thoughts he would not have been able to produce either by himself or in different company. A special mood arises, which would not otherwise affect either partner of the conversation but almost always returns whenever these persons meet again. Prolonged duration of this state produces, from common understanding and mutual misunderstanding, a thought structure [Denkgebilde]* that belongs to neither of them alone but nevertheless is not at all without meaning. Who is its carrier and who its originator? It is neither more nor less than the small collective of two persons. If a third person joins in, a new collective arises. The previous mood will dissolve and with it the special creative force of the former small collective.

We could agree with anybody who calls the thought collective fictitious and the personification of a common result produced by interaction. But what is any personality if not the personification of many different momentary personalities and their common psychological Gestalt? A thought collective, by analogy, is composed of different individuals and also has its special rules of behavior and its special psychological form. As an entity it is even more stable and consistent than the so-called individual, who always consists of contradictory drives.

The individual life of the human spirit contains incongruent elements, such as tenets of faith and superstition which, stemming from various individual complexes, muddy the purity of any theory or system. Both Kepler and Newton, who contributed so much to the modern concept of nature, were ritualistic and religious in their basic attitudes. Rousseau's ideas of education had much greater

---

*Fleck's terminology appears ambiguous, for *Denkgebilde* can designate the structures of thought taken as thought products, mental creations, works of the (communal) mind, etc., but it can also be a synonym for *Denkstruktur* taken as mental structures, thought structures, thought patterns, thought forms, etc. In contexts such as this, both may even be intended in the sense of thought structures created according to thought structures.—Eds.

relevance to the thought collective than to his own individual life.

Every individual belongs to several thought collectives at once. As a research worker he is part of that community with which he works. He may give rise to ideas and developments, often unconsciously, which soon become independent and frequently turn against their originator. As a member of a political party, a social class, a nation, or even a race, he belongs to other collectives. If he should chance to enter some other society, he soon becomes one of its members and obeys its rules. The individual can be examined from the viewpoint of a collective just as well as, conversely, the collective can be considered from that of the individual. Whether in the case of the individual personality or in that of the collective entity, that which specifies the one or the other is accessible only to adequate methods.

The history of science also records cases of independent—one might say personal—exploits. But their independence is only characterized by an absence of collaborators and helpers, or possibly of pioneers; that is, it manifests itself in the personal and independent concentration of historical and contemporary collective influence. In a manner corresponding closely to personal exploits in other areas of society, such scientific exploits can prevail only if they have a seminal effect by being performed at a time when the social conditions are right. The achievement of Vesalius as the originator of modern anatomy was just such an audacious and artistic feat. Had Vesalius lived in the twelfth or thirteenth century he would have made no impact. It is just as difficult to imagine him in that era, as it is, for instance, to imagine Napoleon before the French Revolution. Outside the appropriate social conditions, any development into historical greatness would have been denied to both. The futility of work that is isolated from the spirit of the age is shown strikingly in the case of that great herald of excellent ideas Leonardo da Vinci, who nevertheless left no positive scientific achievement behind.

This by no means implies that the individual must be ruled out as an epistemological factor. His sensory physiology and psychology are certainly very important. But a firm foundation for epistemology cannot be established without investigation of the thought community [*Denkgemeinschaft*]. Let me introduce a somewhat trivial analogy. If the individual may be compared to a soccer player and the thought collective to the soccer team trained for cooperation, then cognition would be the progress of the game. Can an adequate report of this progress be made by examining the individual kicks one by one? The whole game would lose its meaning completely.

The importance of sociological methods in the investigation of intellectual activities was already recognized by Auguste Comte. Recently it was stressed by Durkheim's school in France and by the philosopher Wilhelm Jerusalem among others in Vienna.

Durkheim speaks expressly of the force exerted on the individual by social structures both as objective specific facts and as controlled behavior. He also mentions the superindividual and objective character of ideas belonging to the collective. He describes that which is produced by the activities of the collective intellect, "as we encounter them in language, in religious and magic beliefs, in the existence of invisible powers, and in the innumerable spirits and demons which dominate the entire course of nature and the life of the tribe, and as we meet them in customs and habits."[23]

Lévy-Bruhl, a student of Durkheim, writes: "Ideas belonging to the collective follow laws of their own which, especially in primitive races, we cannot discover by studying the white, adult, and civilized individual. On the contrary, it is the study of those ideas belonging to the collective and their connections in primitive societies that throws some light on the origin of our own categories and logical principles."[24] "This approach will certainly lead to a new and positive epistemology based upon the comparative method."[25] Lévy-Bruhl contests the belief in "the identity of the human mind," "which at all times and in all places is supposed to have remained unchanged as far as logic is concerned."[26] He doubts whether "scientific use can at all be made of the idea of a human mind assumed to be untouched by any experience,"[27] because this concept "is just as chimerical as that of man before society."[28]

Gumplowicz expressed himself very poignantly on the importance of the collective. "The greatest error of individualistic psychology is the assumption that a *person* thinks. This leads to a continual search for the source of thought within the individual himself and for the reasons why he thinks in a particular way and not in any other. Theologians and philosophers contemplate this problem, even offer advice on how one ought to think. But this is a chain of errors. What actually thinks within a person is not the individual himself but his social community. The source of his thinking is not within himself but is to be found in his social environment and in the very social atmosphere he 'breathes.' His mind is structured, and necessarily so, under the influence of this ever-present social environment, and *he cannot think in any other way.*"[29]

Jerusalem dealt with this problem in a number of essays, the last of them bearing the apposite title "Social Conditioning of Thinking and of Thought Patterns." "Kant's firm belief in a timeless, completely immutable logical structure of our reason, a belief that has since become the common heritage of all who adopt an a priori point of view and is maintained with great tenacity also by the latest representatives of this direction of thinking, has not only failed to be confirmed by the results of modern ethnology but proved to be definitely erroneous."[30] "The primitive individual feels himself only a member of his tribe and clings to its traditional way of interpreting sensory perceptions with absolutely incredible tenacity."[31] "I have no doubt, and it is confirmed though the diverse institutions found in primitive societies, that tribesmen reinforce each other's belief in the ubiquity of spirits and demons, which is already sufficient to give these figments of the imagination some degree of reality and stability. This process of mutual corroboration is by no means confined exclusively to primitive societies. It is rather prevalent today, fully effective in our everyday lives. I wish to designate this process and any structure of belief formed and fortified by it *social consolidation.*"[32] "Even particular and objective observations. . . require confirmation by the observation of others. Only then will they become common property and thus suitable for practical utilization. Social consolidation functions actively even in science. This is seen particularly clearly in the resistance which as a rule is encountered by new directions of thought."[33]

All these thinkers trained in sociology and classics, however, no matter how productive their ideas, commit a characteristic error. They exhibit an excessive respect, bordering on pious reverence, for scientific facts.

"When mystical elements lose some of their dominance," writes

Lévy-Bruhl, "objective properties ipso facto attract and retain more attention. The share of perception proper grows in inverse proportion to the withering away of mystical ideas belonging to the collective."[34]

Lévy-Bruhl believes that scientific thought yields concepts which "solely express objective features and conditions of beings and phenomena."[35] But he would find it difficult to define the meaning of "objective features" and "perception proper." Moreover, the attraction of attention by objective properties alleged to occur "ipso facto" is psychologically impossible. The perception of scientifically accepted properties (assuming Lévy-Bruhl considers these "objective") must first be learned. It does not occur ipso facto and, indeed, the ability to perceive scientifically is only slowly acquired and learned. Its prime manifestation is discovery. This occurs in a complex, socially conditioned way, which resembles the origin of other ideas of the collective.

"Once the mentality of primitive societies is opened to experience," Lévy-Bruhl continues, "it also becomes more sensitive to contradiction."[36] "As soon as any society's intellectual structure and institutions develop,...a feeling for, or knowledge of, what physically is possible or impossible emerges and gradually becomes established. The situation is therefore much the same with physical absurdity as it is with logic. The same causes render the prelogical mentality insensitive to both types of absurdity."[37]

We must object in principle that nobody has either a feeling for, or knowledge of, what physically is possible or impossible. What we feel to be an impossibility is actually mere incongruence with our habitual thought style. Until recently the transmutation of elements as well as many other phenomena of modern physics, let alone the wave theory of matter, were regarded as absolutely "impossible." "Experience as such," to which one has or does not have access is chimerical, and thus every being gains experience according to his own fashion. Present experiences are linked with past ones, thereby changing the conditions of future ones. So every being gains "experience" in the sense that he adjusts his way of reacting during his lifetime. Scientific experience in particular derives from special conditions established by the history of ideas and by society. Traditional patterns of training are involved in this experience, which is, however, not accessible to everyone.

Jerusalem, too, believes in the possibility of "thinking purely theoretically" and "stating given facts purely objectively." "Man acquires this ability only slowly and by degrees, to the extent that by conscious effort he overcomes the state of complete social bondage and thus develops into an *independent* and *self-reliant* personality."[38] "Only the strengthened individual acquires the ability to state facts purely objectively and thus learns to think theoretically, that is, free from emotion."[39] Jerusalem calls it the "connection between fact and individual." But how is this to be brought into agreement with his statement previously quoted, about the importance of social consolidation even in science?

"A judgment is now objectively true only when it can be considered as exclusively as possible a *function of the process of judgment.* This new, purely objective criterion of truth, which hitherto in a rather superficial and useless formulation was usually designated "agreement" between judgment and facts, must thus be regarded as a product of the tendency to individualistic development."[40]

To this we must object that any thinking, to be emotionless, must be independent of momentary and personal mood, and flow from the average mood of the collective. The concept of absolutely emotionless thinking is meaningless. There is no emotionless state as such nor pure rationality as such. How could these states be established? There is only agreement or difference between feelings, and the uniform agreement in the emotions of a society is, in its context, called freedom from emotions. This permits a type of thinking that is formal and schematic, and that can be couched in words and sentences and hence communicated without major deformation. The power of establishing independent existences is conceded to it emotively. Such thinking is called rational. The causality relation, for instance, was long regarded as purely rational, yet it was actually a relic of strongly emotive demonological ideas belonging to the collective.[41]

If we attempt a critical separation of the so-called subjective from the so-called objective in concrete terms, we will find again and again the active and passive links within knowledge that were mentioned earlier. Not a single statement can be formulated from passive links alone. Active links, usually inappropriately called "subjective," are always involved. A passive linkage can be considered active from a different point of view, and vice versa, as will be discussed in due course. What is the reason for this special position of current scientific statements as required by the philosophers just quoted?

They believe that our present-day scientific opinions are in complete contrast with all other ways of thinking. As if we had become wise and our eyes had been opened, they believe that we have simply discarded the naive self-consciousness of thought processes which are primitive or archaic. We are supposedly in possession of "correct thinking" and "correct observation," and therefore what we declare to be true *is* ipso facto *true*. What those others such as the primitives, the old people, the mentally ill, or the children declare to be true *seems to be true only to them*. This arch-naive view, which prevents the building up of a scientific epistemology, reminds us very much of the theory of a French philologist of the eighteenth century who claimed that *pain, sitos, bread, Brot, panis* were arbitrary, different descriptions of the same thing. The difference between French and other languages, according to this theory, consisted in the fact that what is called bread in French really was bread.

Natural scientists in their philosophizing commit the opposite and also very typical error. They are aware that there are no "solely objective features and conditions" but only relations governed by a more or less arbitrary reference system. Their error consists in an excessive respect for logic and in regarding logical conclusions with a kind of pious reverence.

To these epistemologists trained in the natural sciences, for instance, the so-called Vienna Circle including Schlick, Carnap, and others, human thinking—construed as an ideal, or thinking as it should be—is something fixed and absolute. An empirical fact, on the other hand, is relative. Conversely, the philosophers previously mentioned with a background in the humanities construe facts as something fixed and human thought as relative. It is characteristic that both parties relegate that which is fixed to the region with which they are unfamiliar.

Would it not be possible to manage entirely without something fixed? Both thinking and facts are changeable, if only because changes in thinking manifest themselves in changed facts. Conversely, fundamentally new facts can be discovered only through new thinking. These points will be taken up later.

The fruitfulness of the thought collective theory is revealed especially in the facility with which it enables us to compare primitive, archaic, naive, and psychotic types of thinking and to investigate them uniformly. It can also be applied to the thinking of a whole nation, a class or any group no matter how it is constituted. I consider the postulate "to maximize experience" the supreme law of scientific thinking. Thus, once the possibility of such comparative epistemology arises, it becomes a duty to carry it out. The old point of view, which is confined to normative pronouncements about "bad" and "good" thinking, is accordingly obsolete.

The views outlined here should not be construed as skepticism. We are certainly capable of knowing a great deal. If we cannot know "everything," according to the traditional position, it is simply because we cannot do much with the term "everything," for every new finding raises at least *one* new problem: namely an investigation of what has just been found. The number of problems to be solved thus becomes infinite and the term "everything" meaningless.

An "ultimate" or set of fundamental first principles from which such findings could be logically constructed is just as nonexistent as this "everything." Knowledge, after all, does not repose upon some substratum. Only through continual movement and interaction can that drive be maintained which yields ideas and truths.

Three

The Wassermann Reaction and Its Discovery

For a long time I wondered how I could describe the Wassermann reaction to a layman. No description can take the place of the idea one acquires after many years of practical experience with the reaction. It is a complex, extremely rich field related to many branches of chemistry, physical chemistry, pathology, and physiology.

The procedure is based on five little-known factors, whose mutual effects are adjusted by means of preliminary tests and whose mode of application is secured through a system of controls. The most important reagent, the so-called "antigen" or, better, "extract," is used on the basis of numerous and varied preliminary tests as well as of comparisons with other previously tested extract preparations. Only a continuous, regular, and well-organized execution of the procedure for the reaction, always with many blood samples, several taken from each series for comparison with the next, will yield results of the necessary reliability. A clinical control of these results must of course also be carried out, involving a comparison of the laboratory results with the clinical results and an appropriate adjustment of the mode of procedure.

Despite every safeguard and mechanization, however, new and unexpected findings continually emerge. From time to time very promising relations and vistas open up, only to vanish again like so many mirages. The reaction occurs according to a fixed scheme, but every laboratory uses its own modified procedure, which is based upon precise quantitative calculations; nevertheless, the experienced eye or the "serological touch" is much more important than calculation. It is possible to obtain a positive Wassermann reaction from a normal blood sample and a negative one from a syphilitic sample without any major technical errors. This was shown very clearly at the Wassermann Congresses held by the League of Nations, where the best serologists from various countries examined the same blood samples simultaneously but independently. It was shown then that the results did not completely agree either with each other or with the clinical aspect of the disease.

Yet the reaction is one of the most important medical aids used in thousands of medical establishments every day and about which many theoretical papers are written. Its importance is already apparent from the fact that the procedure is subject to official regulations and that in many countries only special laboratories are qualified to carry it out.

This field is a little world of its own and therefore can no more be fully described in words than any other field of science. Words as such do not have fixed meanings. They acquire their most proper sense only in some context or field of thought. This delicate shading of the meaning of a word can be perceived only after an "introduction," whether historical or didactic.

But neither approach is purely rational or intellectual per se. History cannot be logically constructed any more than a scientific event, if only because it involves the progress of vague and indefinable concepts which are about to crystallize. The more detailed and differentiated the description is for any field of thought, the more complex, interrelated, and mutually dependent in definition will be its concepts. They become a tangle impossible to unravel logically, an organic structure produced by mutual development and with interacting components. At the end of the process, the beginning cannot be understood any longer or even properly expressed in words. If at all, it will be understood and expressed differently than it was originally. It is therefore not possible to present the result of such a development as if it were a logical conclusion from past premises. Can the development of the concept of chemical elements from the early qualitative concept to the modern one mainly in terms of atomic weight be described in terms of formal logic? The meanings of the concepts of quality, weight, element, composition have changed completely during the course of time, although in harmonious reciprocity. No medieval chemist could understand a modern law of chemistry in the same way that we do today and vice versa.

Nor is the didactic or authoritative type of introduction purely rational, for the current state of knowledge remains vague when history is not considered, just as history remains vague without substantive knowledge about the current state. Any didactic introduction to a field of knowledge passes through a period during which purely dogmatic teaching is dominant. An intellect is prepared for a given field; it is received into a self-contained world and, as it were, initiated. If the initiation has been disseminated for generations as in the case of introducing the basic ideas of physics, it will become so self-evident that the person will completely forget he has ever been initiated, because he will never meet anyone who has not been similarly processed.

One could argue that, if there were such an initiation rite, it would be accepted without criticism only by the novice. The true expert must free himself from the shackles of authority and justify his first principles again and again until he establishes a purely rational system.

But the expert is already a specially molded individual who can no longer escape the bonds of tradition and of the collective; otherwise he would not be an expert. For the introduction, then, factors which are not subject to logical legitimization are also necessary, as well as essential both to the further development of knowledge and to the justification of a branch of knowledge that constitutes a science in itself.

We are now about to perform the rite of initiation into the field of the Wassermann reaction according to the German ritual. I have chosen the 1910 edition of the catechism by Citron, a student of Wassermann. As a textbook it is still rather useful, although already outdated by the most advanced research.

Dr. Julius Citron, *The Methods of Immunodiagnostics and Immunotherapy* (Leipzig 1910). First lecture: Introduction. The concepts of immunity and antibody. The law of specificity. The importance of control experiments.

Gentlemen: There are several approaches we can use to the diagnosis of infectious diseases. Besides clinical observation, which enables us to make the diagnosis through close observation of the temperature curve, changes in the organs, exanthema, and the biochemical processes, we were taught by etiological research to utilize the direct detection of specific causative agents, and by immunology to utilize that of the specific reaction products of the organism, in making the diagnosis. We know now that the progress of an infectious disease depends not only on the type, the quantity, and the virulence of the disease germ, but also on the behavior of the organism. *The disease must be seen from the viewpoint of the reciprocal effect arising from these two groups of factors, although it is impossible to determine in detail the specific effect of the causative agent and its products, and that of the reactive power of the organism.* Although the reaction of the organism varies widely in detail, it can be shown that in spite of all individual differences, well-characterized bacteria and their products are confronted with equally typical basic forms of defense measures serving the organism. For this purpose the body employs cellular and humoral means. It is possible to arrange infectious diseases in an order which shows cellular reactions dominating the picture at one end of the scale, humoral changes at the other, with every intermediate degree between these extremes. We thus find in the widely diversified picture of tuberculosis that the tubercular nodule occurs again and again as a typical cellular reaction product, whereas leprous and syphilitic infections induce the cellular changes that are typical of these diseases. More difficult to recognize, however, because they are invisible both to the naked eye and under the microscope, are those delicate biological reactions which occur in the body fluids during the course of infectious diseases. Special methods are necessary to detect and differentiate humoral changes found especially in the blood serum. But as we now know, humoral immunity reactions, like cellular ones, are not confined to the field of infectious diseases proper but are to a far greater extent expressions of physiological events, whether normal or pathological. With humoral reactions, the ingenious concept of Ehrlich's side-chain theory has enabled us to understand that the physiological manifestation of assimilation which functions in nutrition and energy consumption corresponds to events leading, in pathological conditions, to the formation of the anti-infectious reaction products. Metchnikoff has shown in a no less remarkable analogous achievement that the same group of cells originating in the mesenchyme, which the organism mobilizes against the bacterial enemy, fulfills a variety of physiological and physiological-pathological functions throughout the animal kingdom. They cooperate in the metamorphosis of the body structure of lower animals by their ability to make entire organs disappear. They also take part in the involution of the uterus after childbed, eat up nerve cells destroyed during senile atrophy of the nerve centers, and in the form of chromophages bleach the hair as a sign of advancing age. *The dividing line between the physiological and the pathological event cannot be biologically drawn with any precision.* It represents a whole chain of phenomena with various transitions.

Gentlemen: To make certain that we understand each other in what follows it is above all necessary that we agree on certain concepts, with which most of you are probably already familiar.

To begin with, the word "immunity" needs to be explained. You all know the strange phenomenon that after recovery from most infectious diseases the organism undergoes a change, detectable neither macroscopically, nor microscopically, nor chemically, which protects it against, or at least makes it less susceptible to, the same infectious disease. Because, as you will presently hear, we must distinguish between types of immunity, it is advisable to introduce certain attributes in the interest of facilitating understanding. We designate as "active immunity" the form in which the body immunizes itself by its own power in its fight against infection. You know that Jenner and Pasteur artificially produced this form of immunity, spontaneously acquired through recovery from a disease, for the purpose of protective vaccination and inoculation. Our knowledge about the nature of active immunity is as yet incomplete. All we can do is demonstrate that under active immunity the organism usually forms certain specific reaction products against the disease germs and their toxins. We call these reaction products, which circulate mainly in the blood serum, antibodies. These antibodies have different names according to their different effects, and they vary in significance. The agglutinating and precipitating antibodies, designated as the agglutinins and precipitins respectively, probably have very little protective effect. But others undoubtedly serve to protect the organism either by directly neutralizing bacterial poisons and toxins (antitoxins), by killing bacteria (bacteriolysins, bactericides), or by changing the bacteria in such a way that they can be destroyed more easily by the cells (bacteriotropins, opsonins). Corresponding to these three main types we can speak of antitoxic, bactericidal and cellular immunity, of course with many possible intermediate types. It is very likely that other, still unknown types of immunity exist besides those already known. Above all it can be accepted as certain that cellular immunity can claim far greater importance than is usually accorded it on the basis of the facts known thus far. There is apparently also a type of cell immunity which is effective without the agency of any serum substances, and this is designated "histogenic" immunity and "tissue immunity."

By injecting antibody-containing blood serum obtained from immunized animals into healthy, nonimmunized ones, it is often possible to induce immunity against the associated infective agents. Here the organism thus protected has not produced its protective substances itself through cellular activity of its own but receives them in a prefabricated state. We therefore call this "passive immunity" to distinguish it from the previously discussed "active" form.

All the types of immunity described so far share in the fact that they are acquired only through certain reactions, whether this involves either spontaneous or artificial recovery from disease, or alternatively the transfer of antibodies. Besides such "acquired" immunity, there is also "natural immunity," by which we mean the fact that not every type of animal is susceptible to every infectious disease. Man, for instance, enjoys a natural immunity against a number of the most dreaded animal diseases such as chicken cholera and swine pox. Natural immunity is almost always of the cellular type. The most important natural defensive weapon is phagocytosis, which is the ability of the leucocytes to "eat" the bacteria.

In conclusion it should be pointed out briefly that we speak of a "local" and "general" immunity to express the difference that various organs of the same individual can show in their reaction to an infection. An immunity may be called "relative" or "absolute" to denote quantitative differences, and a further distinction can be made between "permanent" and "temporary" immunity.

Gentlemen: The second important term we must discuss is the concept of the antibody. I have already explained to you briefly that by antibodies we mean all the specific reaction products formed by the organism against disease germs and their products. To complete the picture I must now add that antibodies are also formed when any foreign albumin of a nonbacterial type, for instance blood from a different type of animal or albumin from chicken eggs, is administered to an organism parenterally, that is in a manner other than stomachically.

To establish better understanding of the nature of antibodies, attempts have been made to prepare them in a chemically pure state. All these attempts, however, have thus far failed. The chemical nature of antibodies is unknown. We do not even know whether what we call antibodies constitute independent chemical structures at all. All we know is the serum effects. Thus antibodies represent only the mentally accomplished materialization of these serum effects. But for didactic purposes we shall henceforth speak of different antibodies such as antitoxins or agglutinins when we really mean the antitoxic or the agglutinating ability of the serum.

Although the effectiveness of the individual antibodies differs widely, specificity is a common property of them all. This means that the typhus antibody, for example, can produce the various immunity reactions only with typhus bacteria and the cholera antibody only with cholera vibriones. This property of specificity is so important that we must not designate as antibodies any substances possessing all the other properties of an antibody yet remaining nonspecific. The law of antibody specificity does not apply, of course, in the extreme form I have just outlined to explain the term to you. We shall presently have the opportunity to discuss the nature of specificity in detail, and thus get to know its limitations. *For the time being, however, I would ask you to commit firmly to your memory the law that every true antibody is specific and that all nonspecific substances are not antibodies. The law of specificity is the precondition of serodiagnostics.* Correctly to diagnose typhus, for example, we must know that a patient's serum can produce immunity reactions with genuine typhus bacilli only if the patient in question really has typhus. When the specificity of a reaction becomes doubtful, its diagnostic utilization must accordingly suffer. For this reason, we must repeatedly discuss the question whether and to what extent any given reaction is specific and ascertain true specificity in any way possible, especially by means of control tests. Permit me, even in this first lecture, to draw your attention to the importance of adequate control tests. At first you will find it perhaps pedantic that the controls demanded for seemingly very simple tests often require many times the effort involved in the actual test. You will be tempted perhaps to omit such controls if during the practical utilization of serodiagnostics you are able to obtain good results without the required controls, even in large series of tests. Nevertheless, gentlemen, I cannot impress upon you strongly enough never to operate without the necessary controls. You will thus protect yourselves against grave errors and faulty diagnoses, to which even the most competent investigator may be liable if he fails to carry out adequate controls. This applies above all when you perform independent scientific investigations or seek to assess them. Work done without the controls necessary to eliminate all possible errors, even unlikely ones, permits no scientific conclusions.

I have made it a rule, and would advise you to do the same, to look at the controls listed before you read any new scientific papers dealing with serodiagnostics. If the controls are inadequate, the value of the work will be very poor, irrespective of its substance, because none of the data, although they may be correct, are necessarily so.

What does this excellent introduction suggest? What elements do we find in it that cannot be justified? It will not be difficult to identify them, for we already have the rudiments of other views, even though these have not as yet found their way into the textbooks. The new views, of course, cannot be fully confirmed either, but because the forcefulness of the old views has diminished, we have acquired the possibility of a comparison.

1. *The concept of infectious disease.* This is based on the notion of the organism as a closed unit and of the hostile causative agents invading it. The causative agent produces a bad effect (*attack*). The organism responds with a reaction (*defense*). This results in a conflict, which is taken to be the essence of disease. The whole of immunology is permeated with such primitive images of war. The idea originated in the myth of disease-causing demons that attack man. Such evil spirits became the causative agent; and

the idea of ensuing conflict, culminating in a victory construed as the defeat of that "cause" of disease, is still taught today.

But not a single experimental proof exists that could force an unbiased observer to adopt such an idea. It is unfortunately beyond the scope of our discussion to examine all the phenomena of bacteriology and epidemiology one by one to show that the disease demon haunted the birth of modern concepts of infection and forced itself upon research workers irrespective of all rational considerations. It must suffice here to mention the objections to this idea.

An organism can no longer be construed as a self-contained, independent unit with fixed boundaries, as it was still considered according to the theory of materialism.[1] That concept became much more abstract and fictitious, and its particular meaning depended upon the purpose of the investigation. For the morphologist it has changed into the concept of genotype as the abstract and fictitious result of hereditary factors. In physiology we find the concept of "harmonious life unit," according to Gradmann, "characterized by the notion that the activities of the parts are mutually complementary, mutually dependent upon each other, and form a viable whole through their cooperation." Morphological organisms of the type which are self-contained units do not have this ability. But a lichen, for instance, whose constituents are of completely different origins, one part an alga, another a fungus, constitutes such a harmonious life unit. The constituents are closely interdependent and on their own are usually not viable. All symbioses, for instance, between nitrogen-fixing bacteria and beans, between mycorrhiza and certain forest trees, between animals and photogenic bacteria, and between some wood beetles and fungi form "harmonious life units," as do animal communities such as the ant colony, and ecological units such as a forest. A whole scale of complexes exists which, depending upon the purpose of the investigation, are regarded as biological individuals. For some investigations the cell is considered the individual, for others it is the syncytium, for still others a symbiosis, or, lastly, even an ecological complex. "It is therefore a prejudice to stress the idea of organism," in the old sense of the word, "as a special kind of life unit, a prejudice which is unbecoming to modern biology."[2] In the light of this concept, man appears as a complex to whose harmonious well-being many bacteria, for instance, are absolutely essential. Intestinal flora are needed for metabolism, and many kinds of bacteria living in mucous membranes are required for the normal functioning of these membranes. Some species exhibit for their vital functions an even greater dependence upon others. Their metabolism and propagation, indeed their entire life cycle, depend on a harmonious interference by other species. Some plants are pollinated by certain beetles; and malarial plasmodia depend for their life cycle upon their transmission by mosquito to man.

Now continuous biological changes, within any complex biological individual, so construed are based upon phenomena which can be divided into several categories. They constitute either (1) a kind of spontaneous so-called constitutional process within the genotypes, such as mutations and spontaneous gene changes, roughly comparable with spontaneous radioactive phenomena within an atom. Many a disease belongs to this category, such as the hemolytic icterus of Nägeli, and even the outbreak of certain epidemics might perhaps be included here. Or they are (2) cyclic changes, of which some are genotypically conditioned and others are the result of reciprocal action within the complex life unit. These include the life cycle of organisms (aging), generational change, and some of the dissociation phenomena of bacteria. Both serogenesis and immunogenesis must be listed here, as well as virulence as a life phase of bacteria and even some infectious diseases, such as furunculosis during puberty. Or, lastly, they are (3) pure changes within the constellation of reciprocally acting parts of the unit comparable, for instance, to the reaction among ions in a solution. Hypertrophy of one element of the biological unit at the expense of another is a change of this type, as is the imbalance either consequent upon phenomena of the first or second category, or caused by external physico-chemical conditions. Most infectious diseases belong to this latter class. It is very doubtful whether an invasion in the old sense is possible, involving as it does an interference by completely foreign organisms in natural conditions. A completely foreign organism could find no receptors capable of reaction and thus could not generate a biological process. It is therefore better to speak of a complicated revolution within the complex life unit than of an invasion of it.[3]

This idea is not yet clear, for it belongs to future rather than present biology. It is found in present-day biology only by implication, and has yet to be sorted out in detail.

So construed, the concepts "sickness" and "health" also become unsuitable for any exact application. What used to be called infectious disease or the spread of epidemics belongs partly to the first, partly to the second or even the third group of phenomena. Biologically, this also includes phenomena such as germ carrying, latent infection, the development of allergies, and even serogenesis. These have nothing directly in common with being ill, although they are very important to the mechanism of the disease. The old concept of disease thus becomes quite incommensurable with the new concepts and is not replaced by a completely adequate substitute.

2. *Hence, the concept of immunity in this classical sense must be abandoned.* A fundamental property of all biological events is modified reaction to a repeated stimulus. Sometimes this consists of a certain immunity, whether habituation to toxin, true immunity to disease, or even mechanical immunity such as that against scalding (thickening of the skin) or against bone fracture (callus formation). On other occasions, hypersensitivity occurs, sometimes even in the same cases just mentioned. With sufficiently refined methods it is in fact always possible to detect both together. In some respects there is increased power of resistance and in others increased sensitivity. Thus instead of the prejudicial concept of immunity, we have the general concept of allergy (changed mode of reaction), or according to Hirszfeld the absence of reaction and hyperreactivity. Instead of antibodies, we speak of reagins to stress the lack of direction of the effect, because reagins ensure not only that the irritant is decomposed and rendered harmless but also that it is effective in the first instance and possibly increased in strength or velocity of reaction.

Many classical concepts of immunology were evolved during the period when, under the influence of great chemical successes in physiology, misguided attempts were made to explain the whole, or almost the whole, of biology in terms of effects produced by chemically defined substances. Toxins, amboceptors, and complements were treated as chemical entities, with such adversaries as anti- toxins and anticomplements. This primitive scheme based upon activating and inhibitory substances is being progressively dis- carded in accordance with current physico-chemical and colloidal theories in other fields. We now speak of states or structures rather than substances, to express the possibility that a complex chemico-physico-morphological state is responsible for the changed mode of reaction, instead of chemically defined substances or their mixtures being the cause.

3. *Many other habits of thought* that today cannot be objectively confirmed will also be found in Citron's textbook for serologists.

The division into *humoral* and *cellular* factors (the French stress the second, the Germans the first) cannot be confirmed any more than the concept of specificity in the distinctly mystical sense in which it is used here.

4. Citron's lecture also contains a *methodological initiation.* The novice is introduced as quickly as possible to the importance of "controls." These specifically biological comparison tests, which are to be performed parallel to the main ones, have already been mentioned. There is no universally accepted system of measure- ment in biology, and this is especially so in serology. The results of quantitative tests are read minimetrically with dilution to the limits of reactivity and comparison with standard reagents as well as their combinations. The effect produced by a combination of reagents is also compared with that of incomplete combinations from which just *one* reagent has been intentionally omitted. All these compari- sons control the outcome and are therefore called "controls," Epistemologically it may not be the best method, but we have yet to find another one.

5. The lecture also contains *general precepts* in addition to these particular ones: Cognition should progress not through intuition or from empathy with the phenomena as a whole, but through clinical

and laboratory observation of the various constituent phenomena. The so-called diagnosis—the fitting of a result into a system of distinct disease entities—is the goal, and this assumes that such entities actually exist, and that they are accessible to the analytical method.

Such precepts form the thought style of the serologist's collective. They determine the direction of research and connect it with a specific tradition. It is perfectly natural that these precepts should be subject to continual change. To prevent misunderstandings it must once again be stressed that it is not the purpose of these pronouncements to play off earlier viewpoints against those of today, or those of leading research workers against textbook views. It is altogether unwise to proclaim any such stylized viewpoint, acknowledged and used to advantage by an entire thought collective, as *"truth or error."* Some views advanced knowledge and gave satisfaction. These were overtaken not because they were wrong but because thought develops. Nor will our opinions last forever, because there is probably no end to the possible development of knowledge just as there is probably no limit to the development of other biological forms.

Our sole purpose has been to demonstrate how even specialized knowledge does not simply *increase* but also basically *changes.* Yet we do not want to confine ourselves merely to some banal statement about the transience of human knowledge.

Every act of cognition means that we can first of all determine which passive connections follow of necessity from a given set of active assumptions. To investigate successfully how assumptions change requires research into thought styles. Thought style, suggested during even the earliest acquaintance with any science and extending into the smallest details of its specialized branches, calls for a sociological method in epistemology.

Neither the particular coloration of concepts nor this or that way of relating them constitutes a thought style. It is a definite constraint on thought, and even more; it is the entirety of intellectual preparedness or readiness for one particular way of seeing and acting and no other. The dependence of any scientific fact upon thought style is therefore evident.

Thus even Citron's presentation, which only about twenty years

ago was considered to be at the very zenith of research, indicates a thought collective nexus of knowledge which manifests
itself in a social constraint upon thought. In the course of our
further discussion concerning the Wassermann reaction, this interaction among the individual, the collective, and the fact will be
considered in detail.

If an animal, for instance a rabbit, is inoculated, that is, immunized with killed bacteria or with the blood of a different
species, the serum of the animal in question (the immune serum)
acquires the property of decomposing such bacteria or blood corpuscles. Serologists have, so to speak, materialized this property by
giving the hypothetical, even "symbolic" substance in the immune
serum the name "bacteriolysin" or "hemolysin." Bacteriolysis or
hemolysis succeeds only with *fresh* serum taken from a *pretreated
animal.* If allowed to stand for prolonged periods, or heated to
50-60°C for thirty to thirty-five minutes, the serum will lose this
property, although not irreversibly. Serum deactivated by age or
heat can be reactivated by the addition of *fresh* serum from a *not
pretreated* animal, preferably a guinea pig, even though the latter
serum on its own has no effect whatsoever on those bacteria or
blood corpuscles. It merely supplements the bacteriolysins or
hemolysins of the inactivated immune serum. This property was
also materialized by the serologists. The name "complement" is
given to this hypothetical substance present in the fresh serum and
in whose presence lysis occurs. To induce bacteriolysis or hemolysis
two "substances" are thus necessary: (1) the bacteriolysin or hemolysin, (2) the complement. They act only together. The bacteriolysin and the hemolysin respectively are heat-resistant, that is, they
withstand heating to 56-60°C without damage. The complement is
heat-sensitive. It is lost when heated to 56-60°C as well as during
prolonged storage (aging) of the serum. In the symbolic language
of the German serologists, which owes its origin to Ehrlich, the
antibodies of the bacteriolysin and of the hemolysin type are called
amboceptors, because they combine with and fix two substances:
the one earmarked for immunization, called antigen, and the
other, the complement.

Ehrlich introduced very descriptive and mnemotechnically excellent symbols appropriate for the complex side-chain theory. The

amboceptors are specific; their effect is confined to the particular antigen used in the immunization—being effective only on the blood corpuscles of a ram, only on cholera bacilli, etc. The complement is present in the normal serum and acts with any amboceptor.

It was at one time an open question whether a single uniform complement or several different complements existed in the same normal serum, one complementing the bacteriolysin, and the other the hemolysin. Ehrlich and his followers adopted the pluralistic view, but Bordet and Gengou proved the unitarian view in 1901 with the following experiment. If bacteria (antigen 1) are mixed with the corresponding inactivated immune serum (1) (that is, the bacteriolytic amboceptor), as well as with the complement, bacteriolysis will occur. If one now adds to this a mixture of blood corpuscles (antigen 2) and the corresponding immune serum (2) (that is, the hemolytic amboceptor), no hemolysis will occur, because the complement has been used up in the first process (bacteriolysis) and is no longer available for the second (hemolysis). This can be shown in the symbolic sign language as illustrated.[4]

The complement is completely used up for bacteriolysis and none is left for subsequent hemolysis. This proves that no separate complement exists for hemolysis; that the complement is therefore uniform. The experiment must be conducted quantitatively, of course, which calls for special preliminary tests.

Because it is visible to the naked eye, hemolysis can be detected more easily than bacteriolysis, which requires microscopic examination. This complement fixation method has therefore become the most important instrument in serology, since according to this scheme the hemolytic system (the hemolytic amboceptor plus the corresponding blood corpuscles) can be used to indicate the occurrence of bacteriolysis, that is, whether the bacteriolysin used [is the "specific" one for, and thus]* reacts with the bacteria used. With this method, if the bacteria are known the bacteriolysin can be diagnosed. Conversely if the serum, that is the bacteriolysin, is known, the bacteria can be diagnosed. In the first case we have a method of recognizing, for instance in the serum of patients, the

*The bracketed portion of this sentence is our interpolation.—Eds.

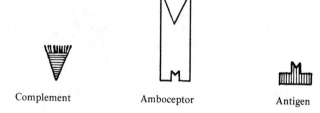

Complement   Amboceptor   Antigen

Fixation of these three substances.

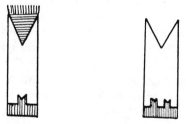

Scheme of the reaction, after Bordet and Gengou 1901. The complement has been used up in the first fixation; no second fixation (hemolysis) is therefore possible.

Figure 2

presence of certain antibodies upon which a diagnosis of the disease can be based. In the second we can determine with very great certainty whether the unknown bacteria belong to the same species as the standard bacteria used for artificial immunization. This complement fixation method according to Bordet and Gengou was soon successfully used by Widal and Le Sourd for abdominal typhus and by Wassermann and Bruck for abdominal typhus and meningitis. Many other workers used it later for such diseases as swine pox, cholera, and gonorrhea.

In 1906 "Wassermann and Bruck proceeded to utilize this reaction for the first time for the detection of antigens in human and animal organ extracts. With the aid of specific tubercle-bacillus immune sera, they demonstrated the presence of lysed tubercle bacillus substances (tuberculin) in tuberculous organs. With the aid of tuberculin, in turn, they demonstrated the occurrence of a specific antibody in the blood, namely antituberculin."[5] These experiments were not rated very highly. Weil expressly wrote of the "untenability of the experiments by Wassermann and his colleague in which specific antigen and antibodies in tuberculous foci, and, in a case of miliary tuberculosis, tubercle bacillus substance in the blood had apparently been successfully demonstrated."[6] Nor did these experiments have any direct major practical or theoretical impact. These results may not have been very solid; nevertheless they were the starting point for Wassermann's syphilis experiments.

It is very interesting to trace the stimulus for these syphilis experiments. Wassermann himself describes the situation as follows: "The head of the Ministry, Friedrich Althoff, asked me to his office when Neisser had returned from his first expedition,* and the French were far ahead in experimental biological research on syphilis. He therefore suggested that I work on this disease to assure that German experimental research have a share in this field."[7] Thus from the very beginning the rise of the Wassermann reaction was not based upon purely scientific factors alone. A

---

*To Indonesia for a suitable climate to conduct his experiments with monkeys.—Eds.

rivalry between nations in a field that even laymen consider very important and a kind of vox populi personified by a ministry official constituted a social motive for the work. The effort expended on this scientific project was correspondingly great. As with the discovery of *Spirochaeta pallida,* here again it was really an organized collective rather than any individual that brought it to completion. Even the lively polemics between, and personal protestations by, the various workers involved, which appeared in the *Berliner Klinische Wochenschrift* during 1921, do not help us to isolate from this community the one, sole discoverer. Owing to the controversy with Ehrlich, the instrument was supplied by Bordet and Gengou. Wassermann and Bruck perfected and expanded it. Because of rivalry with the French, Althoff mapped out the new territory and applied the necessary pressure. Neisser offered the pathological material and his experience as a physician. Wassermann as director of the laboratory was responsible for the plan, and Bruck as his colleague executed it.[8] Siebert prepared the sera. Schucht, an assistant of Neisser's, produced the organ extracts. These are the ones whose names we know. But there certainly were many suggestions concerning technical manipulations, modifications, and combinations from others whom it is impossible to list. Citron decisively improved the dosing. Landsteiner, Marie, and Levaditi, among others, published the first practical method of preparing the extracts. Skills, experience in the field, and ideas whether "wrong" or "right" passed from hand to hand and from brain to brain. These ideas certainly underwent substantive change in passing through any one person's mind, as well as from person to person, because of the difficulty of fully understanding transmitted knowledge. In the end an edifice of knowledge was erected that nobody had really foreseen or intended. Indeed, it stood in opposition to the anticipations and intentions of the individuals who had helped build it. For Wassermann and his co-workers shared a fate in common with Columbus. They were searching for their own "India" and were convinced that they were on the right course, but they unexpectedly discovered a new "America." Nor was this all. Their "voyage" was not straight sailing in a planned direction but an Odyssey with continual change of direction. What they achieved was not even their goal. They wanted evidence for an antigen or an amboceptor. Instead, they fulfilled the ancient wish of the collective: the demonstration of syphilitic blood.

The first paper entitled "A Serodiagnostic Reaction with Syphilis," which appeared on 10 May 1906, was signed by A. Wassermann, A. Neisser and C. Bruck. The purpose of this investigation, as can be gleaned from the contents, was to demonstrate, by means of the complement fixation method, *primarily* antigen in syphilitic organs and in syphilitic blood, and *secondarily* antibodies (cum amboceptors) in the syphilitic blood of patients. The primary aim was pursued with much greater vigor. The authors wrote: "The method consists in taking inactive serum from monkeys pretreated with syphilitic material and mixing it with substances such as organ extracts and serum obtained from syphilitic patients. After the addition of fresh, normal guinea pig serum as the complement, a certain time is allowed for fixation. By means of an inactive, specifically hemolytic serum and its related red blood corpuscles, a test is then performed to show whether the complement first added has been completely or only partially fixed. This manifests itself in the complete or partial failure of lysis of the red blood corpuscles or, in brief, in the degree to which hemolysis is inhibited.[9] It would be of the greatest diagnostic and therapeutic significance if one could succeed regularly in obtaining evidence of syphilitic substances or antibodies in the circulating blood of syphilitics. In a number of cases we have already succeeded in securing this evidence (examining extracts from defibrinated blood instead of the blood serum appears, incidentally, to be more suitable to produce this evidence), but in others we have failed. Obviously the strength of the immune serum has a decisive function here. It must therefore be our next task, which in our climate is perhaps impossible in view of the extreme sensitivity monkeys have in all experiments, to obtain a specific serum of the greatest possible strength against syphilis."[10]

The unbiased observer will consider the reaction described here still very primitive and quite different from what is called the Wassermann reaction today. What then was its most decisive characteristic, immune serum from monkeys, has altogether disappeared, as have extracts from defibrinated blood, because it is not the antigen but only the amboceptors[11] that are required today.

It is important to note that, years later, Bruck, the author of this paper, saw its contents in a light quite different from that of an unbiased observer. He wrote in 1924: "During a discussion between Wassermann, Neisser, and Bruck, the latter was asked to deal with this question. He was able to obtain...positive results and so to demonstrate to Wassermann, his superior at the time, *the original method which remains fundamentally unchanged even today,*[12] and to record it officially. The first communication, entitled 'A Serodiagnostic Reaction with Syphilis,' with Bruck as the author and signed by Wassermann, Neisser, and Bruck, was published at the same time."[13] Retrospectively Bruck saw the ripe fruit already in the seed and hardly noticed that many seeds had not even taken. A similar attitude can be found in Wassermann too.

The second paper by the same authors together with Schucht and entitled "Further Observations on the Demonstration of Specific Syphilitic Substances by Complement Fixation" also appeared in 1906.[14] Evidence of specific syphilitic substances in organ extracts (that is, antigen detection) is again mentioned as being of principal importance, and the search for antibodies in the serum of syphilitics is only of secondary interest. The technique, the necessary controls, and the statistics of the results are each described in detail. Syphilitic antigen was detected in 64 out of 76 extracts from syphilitic organs, including 29 of 29 extracts from confirmed syphilitic fetuses. But not a single one was detected in 7 extracts from brains exhibiting progressive paralysis. Detection of amboceptors —the antibodies—was successful 49 times in 257 samples of syphilitic blood (or 19 percent). This second experimental setup (for amboceptor detection) thus yielded far fewer results than did the first (for antigen detection). It is therefore understandable why the authors should have mentioned antigen detection as of principal importance. Concerning the theory of this reaction, the authors are fully convinced "that it is a specific reaction between syphilitic antigen and syphilitic antibodies"[15] which indicates immunity against spirochaetes. This view was soon supported by the results of Bab and Mühlens which were meant to establish a correlation between the spirochaete count in the livers used for the experiment and the potency of extracts taken from these organs. But support aside, their view was later shown to be in error.

Citron soon showed that the conclusion could not be upheld that the blood corpuscle extracts contained syphilitic antigen, "because such extracts from healthy persons produced the same reaction, although more rarely." Thereafter, such detection of syphilitic antigen was generally rejected, although during the initial experiments it had actually yielded the "good" results and was particularly stressed.

The epistemologically most important turning point occurred with the detection of syphilitic antibodies (amboceptor detection). During the initial experiments it produced barely 15–20 percent positive results in cases of confirmed syphilis. How could it then increase to the 70–90 percent found in later statistics? *This turning point represented the actual invention of the Wassermann reaction as a useful test.* The theory of the reaction as well as the historical and psychological circumstances surrounding its conception are of less practical importance. *If the relation of the Wassermann reaction to syphilis is a fact, it became a fact only because of its extreme utility owing to the high probability of success in concrete cases.* The moment when this decisive turn occurred cannot be accurately determined. No authors can be specified who consciously brought it about. We cannot state exactly when it occurred nor explain logically how it happened.

The turning point has often been discussed. But even the principal actors themselves can say no more than that the technique had first to be worked out. Sometimes Citron is credited with having brought about the turning point through his introducing increased serum dosage. Wassermann and his co-workers originally used 0.1 cc of patient serum, but Citron recommended 0.2 cc. Yet today even 0.04 cc of patient serum is ample, provided all the reagents are mutually adjusted with precision. Fundamentally it is this very reagent-adjustment, coupled with learning how to read the results, that made the Wassermann reaction useful.

Proper balance was difficult to achieve and the results tended to fluctuate. There were too many positive results even with non-syphilitics and too many negative ones even with syphilitics. The optimum intermediate position between minimum nonspecificity and maximum sensitivity had to be gradually established. This, however, is entirely work of a collective consisting mostly of anonymous research workers, adding now "a little more," now "a little less" of a reagent, allowing now "a little longer," now "a little shorter" reaction time, or reading the result "a little more" or "a little less" accurately. Added to this were modifications in the preparation of the reagents and other technical manipulations, such as the controls and preliminary tests as well as titrations and matching. "Some authors," Citron wrote in 1910, "call only those test tubes positive in which complete inhibition of hemolysis has occurred. That this is a poor method is borne out by the statistics published by authors such as Bruck and Stern. A great many definite cases of syphilis are indicated there which react negatively where this extreme criterion is applied, although to all appearances they were positive."[16] This describes the situation in which the sensitivity was insufficient.

Ten years later, in 1921, Weil wrote: "It must be borne in mind in this context that at the time we conducted these experiments the technical development of the Wassermann reaction had not yet been completed. It proceeded in the direction of making the reaction less and less sensitive to obtain a clinically usable test for syphilis. It must also be mentioned that most of the reactions we produced were weakly positive. These were accorded great importance at the time, but later such weak results were no longer considered positive."[17] This describes the situation in which exaggerated sensitivity or nonspecificity was dominant.

Collective experience thus operated in all fields related to the Wassermann reaction until, with disregard for theoretical questions and the ideas of individuals, the reaction became useful. But this rewarding and tedious work of the collective was carried out only as a consequence of the special social importance of the syphilis question and of the problem regarding change in syphilitic blood.

As early as 1907, the many wide-ranging tests had shown that, to produce the antigen (spirochaete substance) required for the reaction, alcohol or aqueous extracts from normal organs could be used unrelated to the specific antigen—that is, to *Spirochaeta pallida*—in place of extracts from confirmed syphilitic organs. Landsteiner, Müller and Pötzl, Porges and Meier, Marie and

Levaditi, Levaditi and Yamanouchi reported this almost simultaneously.

The belief of Wassermann and his co-workers "that a spirochaete antigen and a spirochaete amboceptor, that is, a specific antigen-antibody reaction, had been demonstrated" was therefore completely mistaken. This became all the more obvious after the experiments by Kroó, which proved that no positive Wassermann reaction could be produced in man through immunization with killed spirochaetes, although spirochaete antibodies could be detected. After all, the Wassermann reaction proves only a special change in syphilitic blood, and even today we do not know much more than this. In the place of the antigen conforming to some theory or scheme, alcohol extracts from bovine or human heart are now used almost exclusively. To these, following the suggestion of Sachs, cholesterol may be added.[18] With such extracts syphilis serum produces flocculation, which is clearly visible under certain conditions and on which some special and very practical flocculation reactions are based. The precipitate resulting from the mixture of syphilis serum and organ extract has a special effect, which may be due to adsorption, which removes the complement from the hemolytic system consisting of ram blood corpuscles plus corresponding hemolytic amboceptor. This produces the inhibition of hemolysis, which indicates the positive Wassermann reaction.

According to another theory, namely the autoantibody theory of Weil, the Wassermann reaction is not an instability reaction involving hemolysis as a complex biological indicator, but an immunity reaction with true complement fixation of the Bordet-Gengou type, occurring, however, with decomposed-tissue products of syphilis rather than directly with *Spirochaeta pallida*. The organ extract from healthy persons corresponds to the decomposed-tissue products from patients, which explains its usability. There are other theories too, but, in any case, Wassermann's assumption was wrong.

Bruck himself wrote in 1921 about an "extraordinary stroke of luck" by which "during the practical execution of Wassermann's idea, a syphilis reaction was discovered, the nature of which is still not quite clear today."[19] Weil, also in 1921, claimed that the assumption from which Wassermann proceeded was false but that a discovery of great practical importance was made by accident.[20] Laubenheimer added in 1930: "Although Wassermann and his co-workers were led to discover the method which for short is now called 'the Wassermann reaction' by reasoning which was subsequently proved wrong, the reaction has, during the twenty years of its existence, proved its worth in the diagnosing of syphilis by means of serum, so that even today it cannot be fully replaced by any other more recent method."[21] Plaut lastly comments in 1931 with the wisdom of hindsight. "In view of the current situation respecting serology in general and the Wassermann reaction in particular, some actually wanted to accuse August von Wassermann of having proceeded from false assumptions. If this should really be so—and the case is not yet closed—then it was a blessing that Wassermann did proceed from false assumptions. For had he wanted to wait for the correct ones, he would never have discovered his reaction, because even today, six years after his death, we still do not know the correct preconditions for the reaction. Now and again there have even been foolish suggestions that luck had played a part in the discovery of the Wassermann reaction. In the context of research of this kind we can speak of luck only if the discovery in question is a matter of pure chance. But here exactly the opposite happened. Wassermann found his reaction not by chance but because he looked for it, proceeding quite systematically, naturally on the basis of our then current knowledge. But shrewd ideas are frequently also fortunate ideas, and a skilled hand is often also a lucky hand. Precisely this is an inexplicable part of the nature of a brilliant research scientist's personality who, from the many possible ways to tackle a problem, intuitively chooses the one that leads to success."[22]

It is important to record what Wassermann himself thought about it later. "You will remember that, when I created the serodiagnosis of syphilis, I proceeded from the idea, and with the clear intention, of finding a diagnostically usable amboceptor, that is, a substance which has a fixation relation to an antigen and, after saturation of this affinity, fixes an added complement according to the laws established by Bordet and Ehrlich. With my co-worker C. Bruck I used as antigen the organs of syphilitic patients or of monkeys which Neisser had artificially infected with syphilis."[23]

An impartial judge cannot agree with him even with the best will in the world, because in his first experiments Wassermann was not looking for "a diagnostically usable *amboceptor.*" He was looking primarily for "syphilitic substances" which he thought were "dissolved substances of the micro-organisms," that is, *antigen,* and secondly for "specific antibodies vis-à-vis substances of the causative agent of syphilis," that is, the *specific* amboceptor. But it was later shown that (1) the demonstration of syphilitic substances (antigens) is altogether unsuitable for a diagnostic reaction and (2) the amboceptor indicated by the reaction, if it is an amboceptor at all, is at any rate not a specific amboceptor of the anticausative agent. The ultimate outcome of this research thus differed considerably from that intended. But after fifteen years an identification between results and intentions had taken place in Wassermann's thinking. The meandering progress of development, in all stages of which he was certainly deeply involved, had become a straight, goal-directed path.[24] How could it be otherwise? With the passing of time, Wassermann amassed further experience, and as he did so lost the appreciation of his own errors. It would no longer even be possible for him to "demonstrate 64 times the presence of specific antigen in 69 extracts from syphilitic tissue" and to obtain 14 negative control tests without exception.

The following facts are therefore firmly established and can be regarded as a paradigm of many discoveries. *From false assumptions and irreproducible initial experiments an important discovery has resulted after many errors and detours.* The principal actors in the drama cannot tell us how it happened, for they rationalize and idealize the development. Some among the eyewitnesses talk about a lucky accident, and the well-disposed about the intuition of a genius. It is quite clear that the claims of both parties are of no scientific value. Where a scientific problem is concerned, even one of little significance, these people would not dismiss it so casually. Are we then to maintain that epistemology is no science?

Epistemologically the problem is insoluble from an *individualistic* point of view. If any discovery is to be made accessible to investigation, the *social point of view* must be adopted; that is, the discovery must be regarded as a *social event.*

Early, prescientific ideas brought about a powerful prevailing social attitude toward the problems of syphilis. These were the idea of syphilis as carnal scourge, with strong moralistic connotations;[25] and the persistent idea—demanding justification—of change in syphilitic blood.

The attention, importance, and power of development that this research gained from the special moral emphasis on syphilis cannot be overestimated. For centuries tuberculosis had done far more damage, but it never received comparable attention because, unfortunately, it was not considered the "accursed, disgraceful disease" but often even regarded as the "romantic" one. No tepid rational explanations or statistics can help here. Tuberculosis research simply did not receive as powerful an impulse from society. There was no corresponding social tension seeking relief in research.* The success of our tuberculosis research is therefore not remotely comparable with that of the Wassermann reaction or Salvarsan. Rivalry between nations in the field of pemphigus research would surely be impossible. No head of a public health authority would be able to arouse enthusiasm in the nation's best research workers, because it is a socially unimportant disease. No hospitals, experienced directors, enthusiastic assistants, or public funds could be found. No community discussions, rivalry, or public acclaim would support research. The necessary high tension and feeling for the vital importance of such work would never be generated in a research scientist.

In addition to this prevailing attitude with respect to syphilis, a special one arose from the earlier idea of change in syphilitic blood. Had it not been for the insistent clamor of public opinion for a blood test, the experiments of Wassermann would never have enjoyed the social response that was absolutely essential to the development of the reaction, to its "technical perfection," and to the gathering of collective experience. Wassermann first worked on the serology of tuberculosis. Where then were all those "verifiers," the fortunate fellow-competitors [amici hostes], the countless variations made by caviling rivals? As a result, very little came of this work. Yet surely it was no "worse" than his two first papers on syphilis, which, after

*The first international conference on tuberculosis was held in 1902.—Eds.

all, were also very immature, even if they appeared perfect to the authors and their pupils in the light of their subsequent success.

It was the prevailing social attitude that created the more concentrated thought collective which, through continuous cooperation and mutual interaction among the members, achieved the collective experience and the perfection of the reaction in communal anonymity. The antigen demonstration was rejected, and the initial 15-20 percent of correct results was subsequently increased to 70-90 percent. The findings were stabilized and depersonalized. This thought collective made the Wassermann reaction usable and, with the introduction of the alcohol extract, even practical. It standardized the technical process with genuinely social methods, at least by and large, through conferences, the press, ordinances, and legislative measures.

That which can be explained—where it is assumed that work is exclusively individual—only in terms of accident or miracle, becomes easily understandable where collective work is assumed, as soon as a strong enough motive exists for it. It is an accident when a stone drops into a hole. But it is inevitable that dust should penetrate pores; it is blown about in the environment until it finally enters, but each individual particle comes to rest in its particular position only by accident.

Laboratory practice alone readily explains why alcohol and later acetone should have been tried besides water for extract preparation, and why healthy organs should have been used besides syphilitic ones. Many workers carried out these experiments almost simultaneously, but the *actual authorship is due to the collective, the practice of cooperation and teamwork.*

The problem of how a "true" finding can arise from false assumptions, from vague first experiments, and from many errors and detours, can be clarified by a comparison. How does it come about that all rivers finally reach the sea, in spite of perhaps initially flowing in a wrong direction, taking roundabout ways, and generally meandering? There is no such thing as the *sea as such.* The area at the lowest level, the area where the waters actually collect, is merely *called* the sea! *Provided enough water flows in the rivers and a field of gravity exists, all rivers must finally end up at the sea.* The field of gravity corresponds to the dominant and directing disposition, and water to the work of the entire thought collective.

The momentary direction of each drop is not at all decisive. The result derives from the general direction of gravity.

The genesis and development of the Wassermann reaction can be understood in a similar way. Historically it too appears as the only possible junction of the various trains of thought. The old idea about the blood and the new idea of complement fixation merge in a convergent development with chemical ideas and with the habits they induce to create a fixed point. This in turn is the starting point for new lines everywhere developing and again joining up with others. Nor do the old lines remain unchanged.[26] New junctions are produced time and again and old ones displace one another. This network in continuous fluctuation is called reality or truth.

These last statements must not, however, be taken to mean that the Wassermann reaction can be reconstructed in its objective entirety simply from historical factors along with those of individual and collective psychology. Something inevitable, steadfast, and inexplicable by historical development is always left out of such attempts. It can, for instance, be explained from the collective psychological standpoint that, after the initial work by Wassermann on syphilis serology, many others made it their business both to verify and to "technically perfect" it. The achievement of a positive result and its objective content, however, cannot *in the first instance* be explained through factors of historical development. A very large number of combinations were tried by these "verifiers," but not all were found to be equally good. Only one could be regarded as the best, or at least, only a few could be regarded as good. Which ones are to be so selected cannot be determined from these same factors alone.

The same applies to the problem of the extracts. From the psychological aspect of the collective, it is clear that alcoholic extracts would also be tried besides aqueous ones. But that they are actually suitable cannot in the first instance be explained on the basis of either historical or psychological factors, whether collective or individual. This relates to the problem of active and passive elements in knowledge as broached earlier. The introduction of the alcoholic extract was an active element. Its utility, however, is an inevitable outcome and a passive element with respect to this isolated act of cognition.

We shall presently deal with this problem in greater detail and show that *this compulsion becomes resolved only by comparative epistemological considerations and is explained as an intrinsic constraint imposed by thought style.*

We must first report the historical situation. The early idea of change in syphilitic blood did not cease with the Wassermann stage as described. The Wassermann reaction is far too complex and not clear enough theoretically to have such an effect. The attempts to "replace the complement fixation reaction by other and, if possible, simpler methods are divisible into four large categories. First, attempts were made to produce reactions of both complement fixation and precipitation with the aid of pure lipoids and soaps, whose importance in the serodiagnosis of syphilis became increasingly recognized. In this context we must mention the experiments by Porges-Meier with lecithin, by Sachs-Altmann with cholesterol plus sodium oleate, by Elias, Porges, Neubauer, and Salomon with sodium glycocholate, and by Hermann-Perutz with sodium glycocholate and cholesterol. A second series of experiments concerned the possible practical usability of globulin precipitations. Also in this category are the investigations of Klausner with precipitations by distilled water, as are those of Bruck with precipitations of nitric acid, alcohol, and lactic acid. A third group tried to replace the complement fixation reaction with other chemical and biological methods. The methods introduced by Schürmann (H_2O_2-phenol-ferric chloride), by Landau (iodine oil), and by Wiener-Torday (auric cyanide) among others must be mentioned on the one hand, and those by Weichardt (epiphanin reaction), by Ascoli (meiostagmin reaction), by Karvonen (conglutination), and by Hirszfeld-Klinger (coagulation reaction) on the other. Lastly, with the aid of the organ extracts associated with the complement fixation method, a fourth group of workers attempted the diagnostic utilization of flocculation instead of the complement fixation phenomenon. Here the fundamental investigations by Michaelis, Jacobsthal, and Bruck-Hidaka as well as the methods suggested by Meinicke, Sachs-Georgi, Dold (turbidity reaction), Hecht, Bruck, and others must be mentioned. These reactions must be accorded great practical importance as valuable supplements and controls for the method of complement fixation."[27]

Various modifications and simplifications of the Wasserman

reaction must not be forgotten either. In the so-called active methods of Stern, Noguchi, and others, complement contained in human serum is required instead of that in guinea pig serum. In Bauer's method no hemolytic amboceptor, in the original method obtained from the serum of an immunized rabbit, is added; the one normally found in human serum is used instead. Mutermilch added neither amboceptor nor complement. In yet another method, Sciarra claimed that not even antigen and possibly no addition of extract is necessary, because the antigen is said to be already present in syphilitic blood. There are also a great many modifications concerning the method of inactivating the patient's serum, the use of the complement, the preparation of the extract, the hemolysin production, the mode of using blood corpuscles, and the conservation of the reagents, etc.

The size of the avalanche that the Wassermann reaction set in motion can be estimated from a general paper on the "serodiagnosis of syphilis." In 1927 Laubenheimer cited in it about fifteen hundred papers on this subject, although he restricted himself to more recent work.[28] If foreign-language and little-known contributions are added to these, as well as the clinically-oriented reports, which were not fully considered by Laubenheimer, the number can be estimated today at about ten thousand, including those published since 1927. There certainly cannot be many similar specialized problems which have had so many papers devoted to them.

| Four | **Epistemological Considerations Concerning the History of the Wassermann Reaction** |

1. General Conclusions

If we compare the description of the history of syphilis with that of the Wassermann reaction, we note that the latter requires a much greater number of technical expressions. More basic preparation in the form of greater reliance on expert opinion is necessary, for we are moving away from the world of everyday experience and are entering more deeply into that of scientific specialization. At the same time we are coming into closer contact with the persons involved in such cognition, both collectively and individually. More names must be mentioned.

This is a general phenomenon. The more deeply one enters into a scientific field, the stronger will be the bond with the thought collective and the closer the contact with the scientist. In short, the active elements of knowledge increase.

A parallel shift occurs. The number of passive and inevitable connections produced increases as well, because for every active element of knowledge there corresponds a connection that is passive and inevitable. We have already mentioned a few such linkages, for instance, that the mere use of alcohol in preparing extracts is an active element of knowledge, whereas the actual usefulness of such extracts is a passive one and therefore a necessary consequence.

The same spectacle can be observed in other scientific disciplines. To describe the history of the chemical elements, for instance, we would have to distinguish between two great stages: that of the so-called prescientific theory of the elements and that of scientific chemistry. Active and passive elements of knowledge exist in both. The concepts of the element and of the atom can thus be constructed very effectively from historical factors as well as from those of the thought collective. Such concepts are derived, one might say, from the collective imagination. But the usefulness of these concepts in chemistry is a circumstance which is really independent of any individual knower. The origin of the number 16 for the atomic weight of oxygen is almost consciously conventional and arbitrary. But if 16 is assumed as the atomic weight for O, oxygen, of necessity the atomic weight of H, hydrogen, will inevitably be 1.008. This means that the ratio of the two weights is a passive element of knowledge.

The situation we want to demonstrate consists in the fact that, during the first stage of its history, both the active and the passive elements of knowledge are smaller in number than in the second. Every rule and every chemical law can be divided into an active and a passive part. The more deeply we penetrate into a field, the greater will be the number of *both parts* and not just of the passive ones as might be expected at first glance.

For the time being we can define a scientific fact as *a thought-stylized conceptual relation which can be investigated from the point of view of history and from that of psychology, both individual and collective, but which cannot be substantively reconstructed in toto simply from these points of view.* This expresses the inseparable relation between active and passive parts of knowledge as well as the phenomenon that the number of both these parts of knowledge increases with the number of facts.

Another phenomenon must be noted. The more developed and detailed a branch of knowledge becomes, the smaller are the differences of opinion. In the history of the concept of syphilis we encountered very divergent views. There were far fewer differences during the history of the Wassermann reaction, and as the reaction develops further, they will become even rarer. It is as if with the increase of the number of junction points, according to our image of a network (on page 79), free space were reduced. It is as if more resistance were generated, and the free unfolding of ideas were restricted. This is very important, though it belongs no longer to the analysis of fact but to the analysis of error.

2. Observation, Experiment, Experience

Observation and experiment are subject to a very popular myth. The knower is seen as a kind of conquerer, like Julius Caesar winning his battles according to the formula "I came, I saw, I conquered." A person wants to know something, so he makes his observation or experiment and then he knows. Even research workers who have won many a scientific battle may believe this naive story when looking at their own work in retrospect.

At most they will admit that the first observation may have been a little imprecise, whereas the second and third were "adjusted to the facts." But the situation is not so simple, except in certain very limited fields, such as present-day mechanics, in which there are very ancient and widely known everyday facts to draw upon. In more modern, more remote, and still complicated fields, in which it is important first of all to learn to observe and ask questions properly, this situation does not obtain—and perhaps never does, originally, in any field—until tradition, education, and familiarity have produced *a readiness for stylized* (*that is, directed and restricted*) *perception and action;* until an answer becomes largely pre-formed in the question, and a decision is confined merely to "yes" or "no," or perhaps to a numerical determination; until methods and apparatus automatically carry out the greatest part of our mental work for us.

Wassermann and his co-workers experimented according to the method of Bordet-Gengou, trying to detect the presence of the syphilitic antigen in organ extracts and of syphilitic antibodies in the blood. From the early work we glean far more of hope than of concrete results. Successful experiments are discussed along with those that were unsuccessful, without the reason for failure being accurately known to the authors. It is certain that they were on the wrong track concerning the significance of the titration level with the immune serum from monkeys. In the second experiment the number of successful tests, which means those yielding the expected result, had already risen sufficiently for statistics to be published. Of 76 extracts from syphilitic organs, the syphilis antigen was detected in 64 cases. Of the 76, 7 were from progressive-paralytic brains, all of which were unsuccessful, and Weil had his own ideas about this. If these 7 cases using brain extracts are ignored, the success rate is almost 93 percent. All 14 control tests with confirmed nonsyphilitic extracts were negative; that is, they conformed 100 percent to expectations.

But today we know that such results are beyond all reasonable expectations. First, antigen detection in organ extracts is difficult, and even with the best technique yields only very irregular results. Second, extracts from organs which are definitely nonsyphilitic can also fix the complement with syphilis serum. The control tests with negative results are therefore unintelligible, and the high percentage of positive results is very fortuitous. At any rate, the first experiments by Wassermann are irreproducible.

His basic assumptions were untenable, and his initial experiments irreproducible, yet both were of enormous heuristic value. This is the case with all really valuable experiments. They are all of them uncertain, incomplete, and unique. And when experiments become certain, precise, and reproducible at any time, they no longer are necessary for research purposes proper but function only for demonstration or ad hoc determinations. To understand Wassermann's first experiments, we must imagine ourselves in his position. He had a complete plan and felt certain of the result. But the method was still very crude. It seriously disturbed him, for instance, that he had to use human syphilis material for the immunization of most of his monkeys, since pure cultures of *Spirochaeta pallida* could not yet be produced at the time. There were of course control animals which were inoculated with monkey material. But quite a large number of his monkeys yielded a serum which in addition to syphilis antibodies also contained antibodies against human albumin. The complement fixation with this serum was therefore not always specific to syphilis. Furthermore, titration of the extracts and all other preliminary experiments had not yet been perfected. Hence, the reagents were not yet precisely matched. Moreover, it was not yet known what degree of hemolysis inhibition was to be regarded as positive and what as still negative (see chap.

3 at notes 16 and 17). It is therefore clear that the indicators of the experiments were not well defined. The results of some were ambiguous, and it often had to be decided whether the result of an experiment should be considered positive or negative. It is also clear that from these confused notes Wassermann heard the tune that hummed in his mind but was not audible to those not involved.[1] He and his co-workers listened and "tuned" their "sets" until these became selective. The melody could then be heard even by unbiased persons who were not involved. Who could define the moment when this became possible for the first time? The community of those who made the tune audible and of those who listened increased steadily. It is not appropriate to speak of either correctness or incorrectness in these first experiments, because something very correct developed directly from them, although the experiments themselves could not be called correct.

If a research experiment were well defined, it would be altogether unnecessary to perform it. For the experimental arrangements to be well defined, the outcome must be known in advance; otherwise the procedure cannot be limited and purposeful. The more unknowns there are and the newer a field of research is, the less well defined are the experiments. Once a field has been sufficiently worked over so that the possible conclusions are more or less limited to existence or nonexistence, and perhaps to quantitative determination, the experiments will become increasingly better defined. But they will no longer be independent, because *they are carried along by a system of earlier experiments and decisions,* which is generally the situation in physics and chemistry today. Such a system could then become a self-evident law unto itself. We would no longer be aware of its application and effect. And if after years we were to look back upon a field we have worked in, we could no longer see or understand the difficulties present in that creative work. The actual course of development becomes rationalized and schematized. We project the results into our intentions; but how could it be any different? We can no longer express the previously incomplete thoughts with these now finished concepts.

Cognition modifies the knower so as to adapt him harmoniously to his acquired knowledge. This situation ensures harmony within the dominant view about the origin of knowledge. Whence arises the "I came, I saw, I conquered" epistemology, possibly supplemented by a mystical epistemology of intuition.

This exemplifies the effect of the harmony of illusions (or, as we can now call it, the intrinsic harmony of thought style), which makes the scientific results applicable and generates a firm belief in a reality existing independently of us. Rational epistemology, however, is based upon the acceptance of the *threefold function of cognition* and the reciprocal relations between cognition and its three factors. It necessarily leads to the investigation of thought style as its proper object.

Our remarks about experiment apply to an even greater degree to observation, for experiment is observation directed in a certain way. Let us consider some observations which I recently published in the area of bacterial variability. These were new to me, at any rate.[2]

We grew a streptococcus from the urine of a female patient. Its unusually rapid and profuse growth attracted our attention, as did pigment formation, which is very rare with streptococci. I had never seen streptococci producing such intense pigment and remembered only vaguely having read about them. I therefore wanted to find out about the germ in greater detail. I had intended to grow regular nutrient cultures and perform animal inoculations, as well as a few serological experiments and especially a chemical analysis of the pigment. But the project turned largely into a study of variability. How could this have happened?

A few months previously, at the request of some colleagues, I had prepared a comprehensive survey on the concept of species in bacteriology, which brought me into close contact with the phenomena of variability in bacteria. The colityphus group, difficult to systematize because of its special variability, particularly attracted my attention. I collected details about such factors as mutation, habitat modifications, and so-called germ transmission and saw that without order in the field of variability no consistent concept of species would be possible. Such order, however, could not be established without a fundamental discussion of the concept of the individual, which brought me into contact with the relevant work of Van Loghem's school.

This was the psychological foundation for the observations on streptococcus. Now streptococcus habitually reminds laboratory scientists of staphylococcus. I remembered having read of the splitting off of staphylococcus colonies of different colors. I therefore suggested to my colleague that she find out whether our strain split into lighter and darker colonies. I received the answer the next day. Such a dissociation had just occurred. In addition to the hundreds of ordinary yellowish, transparent colonies, a few very small, white, and more opaque ones had grown. We next carried out an entire series of experiments with several generations of the streptococcus to determine: (1) whether the few small colonies belonged to our strain, (2) the extent to which these differed from the others.

The answer to the first question was positive because these colonies contained organisms that were morphologically, biochemically, and zoopathologically identical with those of the typical colonies. The second part of the investigation called both for many exploratory tests to select the method and for many reformulations of the problem. We could not even claim with any certainty and assurance that a real problem existed at all. Were the new colonies definitely different from the old ones? Differences noticed initially, such as the small size, the lighter color, and the opacity all became unstable in subsequent generations. Strangely, however, a difference remained which at first could not even be clearly understood—the difference between the offspring of the special colonies and that of the others. Not only did it persist, but it in fact increased with the transfers, by the partly subconscious selection of the most divergent colonies during inoculation. All attempts to formulate this difference had to be dropped right after the next reinoculations; until at last, after we had gained comprehensive experience, a formulation crystalized. We were dealing with splitting off not of variants more strongly or weakly pigmented but of colonies with a different structure, although of the same color. In other words, the structural variations of the colonies were much more marked than those of color intensity. Moreover, structural variants were produced which, unlike the color variants, could be perpetuated through transfers. Inoculation of these different colonies finally produced what we later called the smooth type (type G) of streptococcus colony in contrast to the curly type (type L).

The smooth types arising later were always more transparent than the curly ones. The more opaque colonies, which were noted in the initial observations on dissociation and which formed the starting point of the investigation, were therefore not identical with them. Was it, then, a dissociation phenomenon at all? This question must remain undecided, for our first observations are irreproducible. We cannot even describe them clearly, because the descriptive terms and concepts which developed during the work are inadequate for unconditioned observation.

This description of our limited experiment with streptococci can serve as an epistemological example. It shows (1) the material offering itself by accident; (2) the psychological mood determining the direction of the investigation; (3) the associations motivated by collective psychology, that is, professional habits; (4) the irreproducible "initial" observation, which cannot be clearly seen in retrospect, constituting *a chaos;* (5) the slow and laborious revelation and awareness of "what one actually sees" or *the gaining of experience;* (6) that what has been revealed and concisely summarized in a scientific statement is an artificial structure, related but only genetically so, both to the original intention and to the substance of the "first" observation. The original observation need not even belong to the same class as that of the facts it led toward.

Consequently it is all but impossible to make any protocol statements [*Protokollsätze*] based on direct observation and from which the results should follow as logical conclusions. This can be done only during the subsequent confirmation of a finding [*eines Wissens*] but not while making the effort of acquiring it. The results can be no more expressed in the language of the initial observations than, vice versa, the first observations in the language of the results.

Every statement about "First Observations" is an assumption. If we do not want to make any assumption, and only jot down a question mark, even this is an assumption of questionability, which places the matter in the class of scientific problems. This is also a thought-stylized assumption.

One might think that the statement, "Today one hundred large, yellowish, transparent and two smaller, lighter, more opaque colonies have appeared on the agar plate," could in our case be regarded as a description purely of what is observed, devoid of any assumptions. But the statement contains much more than "pure observation" and much more than could in the first instance be claimed with certainty. It anticipates a difference between the colonies, which could actually be established only at a later stage of a long series of experiments. The difference of course—and this is very important—was ascertained as of quite another kind than that anticipated.

No two completely identical colonies were found. We therefore had 102 differently structured colonies. First of all it was necessary to determine whether this or that difference was *important enough* to enable us even to speak of different colonies, and whether such a distinction was scientifically worthwhile. We still had to determine *whether* and *how* common *types of colony* could be established from such different colonies. That these two colonies could constitute something different from the other hundred, and that they somehow belonged together, was not "pure observation" but already a hypothesis, which may or may not prove to be true or, alternatively, from which another hypothesis may evolve.

For all practical purposes, the knower is initially unaware of the hypothetical nature of his assertion. Although the statement mentioned here does not describe a "pure observation," it might well be taken to express a "direct observation" or what a *trained person* would see without difficulty when looking at our agar plate. An expert or specialist in variability phenomena of bacteria, for example, would not be in the least misled by the various forms of all the colonies. He would not stop at "unimportant differences" but would recognize the two types of colony at first glance, without any analysis or hypothesis.

One could, however, argue that, although a "pure observation, that is, one without assumptions" does not occur psychologically, it is logically possible and even necessary as a subsequent construction for the legitimation of a finding. Specifically in our case, such an expert would immediately identify the two different colonies among the 102 but neglect the accidental and unimportant differences among the other 100. This ability, acquired through experience, of immediately drawing a conclusion, during observation, from a long series of comparisons and combinations could, and in fact must, also be carried though very strictly and in detail. The corresponding procedure would be to investigate *all* 102 colonies as to *all* their properties and their theoretically possible combinations and in this way to find the various types of colonies according to their *complete* nature. This is what one might find:

I.	Colonies of 5-6 mm diameter		30
	4-5 mm	"	60
	3-4 mm	"	10
	½-1mm	"	2
			102

II.	Colonies of color 100 (arbitrary scale)		70
	" 80 (lighter)		25
	" 70 "		5
	" 5 "		2
			102

Then the procedure would be repeated for transparency and for *all* other properties. If one were to compare the data in the two tables with each other and to place the relevant colonies beside one another, tabulated according to their ranking, one would find that very light color, together with other conspicuous properties, occurs only in the two very small colonies. Furthermore, the differences between these two colonies and all the others far exceed the fluctuations among the properties of the others when they are compared with one another. They would therefore constitute a distinct type of colony, which was the point to be demonstrated and which would thus have been demonstrated without any assumptions having been made.

This description contains some gross errors, which are committed by many theoreticians. *First,* assumptions are already incorporated within the choice and limitation of the object of investigation. With 102 undoubted colonies, there are certain to be a few doubtful features such as grains or dots that might be regarded as colonies or even as accidental structures, depending upon the assumptions.

Second, it is altogether pointless to speak of *all* the characteristics of a structure. The number of characteristics can be as large as desired, and the number of *possible determinations of characteristics* depends upon the habits of thought of the given scientific discipline; that is, it already contains directional assumptions. Accordingly such mechanical combinatorial analyses are either arbitrary or actually conditioned by thought style.

Third, new discoveries cannot be carried out by such tabulations and mechnically exhaustive combinations any more than, for instance, a poem can be composed by means of combining letters mechanically.

Observation without assumption,[3] which psychologically is nonsense and logically a game, can therefore be dismissed. But two types of observation, with variations along a transitional scale, appear definitely worth investigating: (1) *the vague initial visual perception,* and (2) *the developed direct visual perception of a form.*

Direct perception of form [*Gestaltsehen*] requires being experienced in the relevant field of thought. The ability directly to perceive meaning, form, and self-contained unity is acquired only after much experience, perhaps with preliminary training. At the same time, of course, we lose the ability to see something that contradicts the form. But it is just this readiness for directed perception that is the main constituent of thought style. Visual perception of form therefore becomes a definite function of thought style. The concept of being experienced, with its hidden irrationality, acquires fundamental epistemological importance, which will presently be discussed in detail.

By contrast, the vague, initial visual perception is unstyled. Confused partial themes in various styles are chaotically thrown together. Contradictory moods have a random influence upon undirected vision. There is a rivalry among visual fields of thought. Nothing is factual or fixed. Things can be seen almost arbitrarily in this light or that. There is neither support, nor constraint, nor resistance and there is no "firm ground of facts."

All empirical discovery can therefore be construed as a supplement, development, or transformation of the thought style.

Why did bacteriologists for a time almost fail to see the phenomena of variability? At first there was a period of controversy, involving unconnected details, when variability was too much taken for granted. Billroth, for instance, firmly believed in a universal coccobacterium septicum, which could transform itself into all possible forms. This was followed by the classical Pasteur-Koch period. The all-persuasive power of practical success and personalities created a rigid thought style in bacteriology. Only a strictly orthodox method was recognized, and the findings were accordingly very restricted and uniform. For example, cultures were reinoculated generally for only twenty-four hours. Very fresh cultures (two to three hours) or very old (about six months) ones were not even considered worth examining. As a result, all secondary changes in the cultures, which were the starting point for the restyled theory of variability, escaped attention. Whatever failed to conform completely to the standard scheme was regarded as a "form of involution," a kind of pathological phenomenon, or an "artificial" modification caused by external conditions. The harmony of illusions was thus preserved. Species were fixed, because a fixed and restricted method was applied to the investigation. The thought style, developed in this particular way, made possible the perception of many forms as well as the establishment of many applicable facts. But it also rendered the recognition of other forms and other facts impossible. Now things are turning around. The notion of variability was never quite extinct, but the successors of the classical school regarded any such observations as technical mistakes to be simply passed over in silence or rejected. The first detailed observation of variation to be taken somewhat seriously was made in 1906 by Neisser and Massini. This concerned the so-called bacterium *Coli mutabile*. It could not very well be suppressed, because it was couched throughout in terms of the current thought style and was expressly revolutionary in only one point. The authors used the classical method with only a *single* modification. They examined* the cultures not only after twenty-four hours but again after several days. Had they introduced several modifications all at once, they would have had to wait much longer for a consideration of their findings. They found that after a few

---

*"Examined" here renders *untersucht,* but further reinoculation (*Umimpfen*) was presumably involved; otherwise the desired contrast with traditional method is lacking.—Eds.

days buds containing modified germs were growing within the colony. Reinoculation of these buds and with them also other secondary growth phenomena within the bacterial colonies soon became popular topics for investigation. The spell cast by the harmony of illusions was thus broken, and the conditions were created without which many discoveries would have been impossible. It is typical that the new theory of variability found roots in a country other than that of classical bacteriology. It thrived in America with its paucity of tradition and was attacked most strongly in Koch's native country. It is also typical that this did not constitute a simple regression to the age of transformation of species. The very concept of species as well as many other concepts now became construed in a manner different from that in the past. What is involved here is neither mere accretion of knowledge nor a simple link-up with the period before Koch, but a change in thought style. It is also characteristic that during this change in thought style, or learning by experience, the observation of Neisser and Massini, which was its first stimulus, remained outside the new field. Today it is not considered "classical" variability (the word "classical" can already be used in such a context) but as a bacteriophage effect.

This example also exhibits three stages: (1) vague visual perception and inadequate initial observation; (2) an irrational, concept-forming, and style-converting state of experience; (3) developed, reproducible, and stylized visual perception of form.

This description demonstrates how a finding originates. Many a research scientist will certainly recognize an analogy here with his own method of research. The first, chaotically styled observation resembles a chaos of feeling: amazement, a searching for similarities, trial by experiment, retraction as well as hope and disappointment. Feeling, will, and intellect all function together as an indivisible unit. The research worker gropes but everything recedes, and nowhere is there a firm support. Everything seems to be an artificial effect inspired by his own personal will. Every formulation melts away at the next test. He looks for that resistance and thought constraint in the face of which he could feel passive. Aids appear in the form of memory and education. At the moment of scientific genesis, the research worker personifies the totality of his physical and intellectual ancestors and of all his friends and enemies. They both promote and inhibit his search. The work of the research scientist means that in the complex confusion and chaos which he faces, he must distinguish that which obeys his will from that which arises spontaneously and opposes it. This is the firm ground that he, as representative of the thought collective, continuously seeks. These are the passive connections, as we have called them. The general aim of intellectual work is therefore maximum *thought constraint with minimum thought caprice.*

This is how *a fact* arises. *At first there is a signal of resistance in the chaotic initial thinking, then a definite thought constraint, and finally a form to be directly perceived.* A fact always occurs in the context of the history of thought and is always the result of a definite thought style.[4]

It is the aim of all empirical sciences to establish this "firm basis of facts." Two points are important in epistemology. *First,* this work is continuous. It has no demonstrable beginning and is open-ended. Knowledge exists in the collective and is continually being revised. The store of facts also changes. What has previously been classed with the passive elements of knowledge may later join the active ones. The ratio between the atomic weight of oxygen and that of hydrogen, 16:1.008, for instance, we explained as a proportion resulting passively under given conditions. If, for instance, it were possible to split O into two elements, this proportion would be accounted for by the inadequacy of the earlier method and would have to be replaced by another ratio.

Second, however, it is impossible to exhibit the passive elements of knowledge on their own, as has already been pointed out.

The passive and the active elements cannot be separated from each other completely either logically or historically. Indeed, it is not even possible to invent a fairy tale which does not contain some inevitable connections. Myth differs from science in this respect only in style. Science seeks to include in its system a maximum of those passive elements *irrespective* of inherent lucidity. Myth contains only a few such passive elements, but they are artistically composed.

The necessity of being experienced introduces into knowledge an

irrational element, which cannot be logically justified. Introduction to a field of knowledge is a kind of initiation that is
performed by others. It opens the door. But it is individual experience, which can only be acquired personally, that yields the capacity
for active and independent cognition. The inexperienced individual
merely learns but does not discern.

Every experimental scientist knows just how little a single experiment can prove or convince. To establish proof, an entire
system of experiments and controls is needed, set up according to
an assumption or style and performed by an expert. The state of
being experienced [*Erfahrenheit*], as it will here be designated,
consists in just such factors as (1) the ability to make assumptions
and (2) both manual and mental practice together with a research
scientist's entire experimental and nonexperimental fund of knowledge, including features clearly conceived, those that are uncertain, and those that are "instinctive." The summarized report
about a field of research always contains only a very small part of
the worker's relevant experience, and not even the most important.
Missing is that which makes the stylized visual perception of form
possible. It is as if the words of a song were published without the
tune.

Wassermann's reports about his reaction contain only the description of the relation between syphilis and a property of the
blood. But this is not the most important element. What is crucial
is the *experience* acquired by him, by his pupils and in turn by
theirs, in the practical application and effectiveness of serology.
Without this *experience* both the Wassermann reaction and many
other serological methods *would not have become reproducible and
practical.* Such a *state of experience* became general only slowly
and had to be practically acquired by each initiated individual. A
state of this kind is what the first critics of the Wassermann
reaction lacked. The roots of this state in Wassermann and his
co-workers have already been described. But, even today, anybody
performing the Wassermann reaction on his own must first have
acquired comprehensive experience before he can obtain reliable
results. Only through this experience will he participate in the
thought style, and it is experience alone that enables him to perceive the relation between syphilis and blood as a definite form.

We might also mention some cases where such experience involving the irrational "serological touch" is specifically needed.

1. The preparation and titration of the organ extracts perhaps calls most for experience. Here the need is not confined to theory but includes the skill of preparing uniform dilutions of the extract. An inexperienced individual obtains irregular results through having diluted the extract either too rapidly or too slowly. In this respect the Wassermann reaction is particularly sensitive. It can be confirmed now and again that the kind of extract dilution determined by a given individual does not always automatically work with another person. Psychological and physical differences among the performers of this serological test lead to appreciable differences in the degree to which the colloidal solution from the alcoholic extract disperses. The solution must thus be freshly prepared for each test.

2. The matching of all the five required reagents, so as to maximize the effect of the reactions and ensure that the results are as clear as possible, requires experience. Even quasi-*orchestral practice* is needed if, as is usual, the test is performed by a team. Change in personnel often produces a disturbance in the progress of the reaction, even if the new member of the team had worked well with other associates. This explains the poor results obtained even by excellent research workers at the previously mentioned Wassermann conferences held under the auspices of the League of Nations.

3. Obviously, general competence is also necessary in the elementary operations such as measuring, pipetting, storing of the sera, washing of the vessels, etc.

We can summarize as follows our theory of the recognition of the relation between the Wassermann reaction and syphilis. The discovery—or the invention—of the Wassermann reaction occurred during a unique historical process, which can be neither reproduced by experiment nor confirmed by logic. The reaction was worked out, in spite of many errors, through socio-psychological motives and a kind of collective experience. *From this point of view the relation between the Wassermann reaction and syphilis—an undoubted fact—becomes an event in the history of thought.* This fact cannot be proved *with an isolated experiment* but only with broadly based experience; that is, by *a special thought style* built up from earlier knowledge, from many successful and unsuccessful experiments, from much practice and training, and—epistemologically most important—from *several adaptations and transformations of concepts.* Without this experience the concept of syphilis and that of serum reaction could not have been established and research workers could not have been *trained* to practice accordingly. Error and the failure of many experiments are also part of the building materials for a scientific fact. The perfection of the Wassermann reaction can be seen from this point of view as the solution to the following problem: How does one define *syphilis* and set up a *blood test,* so that after *some experience* almost any research worker will be able to demonstrate a *relation between them* to a degree that is adequate in practice? The collective character of this finding readily manifests itself in such a formulation of the problem; it is based on the need to obtain indispensable experience by comparing working methods with those of other workers, as well as on the need for some kind of connection with the traditional and incomplete concept of syphilis and that of the blood test.

The factuality of the relation between syphilis and the Wassermann reaction consists in just this kind of solution to the problem of minimizing thought caprice, under given conditions, *while maximizing thought constraint.* The *fact* thus represents a *stylized signal of resistance in thinking.* Because the thought style is carried by the thought collective, this "fact" can be designated in brief as the *signal of resistance by the thought collective* [*denkkollektives Widerstandsaviso*].

3.   Further Observations Concerning Thought Collectives

The preceding chapter tried to show how even the simplest observation is conditioned by thought style and is thus tied to a community of thought. I therefore called thinking a supremely social activity which cannot by any means be completely localized within the confines of the individual.

Teamwork can take two forms. It can be simply additive, as when a number of people join together to lift something heavy. Alternatively it can be collective work proper—not merely the summation of individual work but the coming into existence of a special form, comparable to a soccer match, a conversation, or the playing of an orchestra. Both forms occur in thinking and especially in the act of cognition. How could the performance of an orchestra be regarded as the work only of individual instruments, without allowance for the meaning and rules of cooperation? It is just such rules that the thought style holds for thinking. *All paths toward a positive, fruitful epistemology lead toward the concept of thought style,* the varieties of which are mutually comparable and can each be investigated as a result of historical development.

Like any style, the thought style also consists of a certain mood and of the performance by which it is realized. A mood has two closely connected aspects: readiness both for selective feeling and for correspondingly directed action. It creates the expressions appropriate to it, such as religion, science, art, customs, or war, depending in each case on the prevalence of certain collective motives and the collective means applied. We can therefore *define thought style as* [the readiness for] *directed perception, with corresponding mental and objective assimilation of what has been so perceived.* It is characterized by common features in the problems of interest to a thought collective, by the judgment which the thought collective considers evident, and by the methods which it applies as a means of cognition. The thought style may also be accompanied by a technical and literary style characteristic of the given system of knowledge.

Because it belongs to a community, the thought style of the collective undergoes social reinforcement, as will shortly be discussed. Such reinforcement is a feature of all social structures. The thought style is subject to independent development for generations. It constrains the individual by determining "what can be thought in no other way." Whole eras will then be ruled by this thought constraint. Heretics who do not share this collective mood and are rated as criminals by the collective will be burned at the stake until a different mood creates a different thought style and different valuation.

But every thought style leaves remnants. First, there are the small, isolated communes which adhere unchanged to the old style. This explains the existence even today of astrologers and magicians: eccentrics who associate with the uneducated of the lower social classes or become charlatans because they do not share the community mood. Second, every thought style contains vestiges of the historical, evolutionary development of various elements from another style. Probably only very few completely new concepts are formed without any relation whatsoever to earlier thought styles. It is usually only their coloring that changes. Just as the scientific concept of force originated from the everyday concept of force, so also the new concept of syphilis descended from the mystical.

A historical connection thus arises between thought styles. In the development of ideas, primitive pre-ideas often lead continuously to modern scientific concepts. Because such ideational developments form multiple ties with one another and are always related to the entire fund of knowledge of the thought collective, their actual expression in each particular case receives the imprint of uniqueness characteristic of a historic event. It is, for instance, possible to trace the development of the idea of an infectious disease from a primitive belief in demons, through the idea of a disease miasma, to the theory of the pathogenic agent. As we have already hinted, even this latter theory is already close to extinction. But while it lasted, only one solution to any given problem conformed to that style. (See chap. 2, sec. 4, on Schaudinn's "causative agent" versus that of Siegel.) *Such a stylized solution, and there is always only one, is called truth.* Truth is not "relative" and certainly not "subjective" in the popular sense of the word. It is always, or almost always, completely determined within a thought style. One can never say that the same thought is true for A and false for B. If A and B belong to the same thought collective, the thought will be either true or false for both. But if they belong to different thought collectives, it will just *not* be *the same* thought! It must either be unclear to, or be understood differently by, one of them. Truth is not a convention, *but rather (1) in historical perspective, an event in the history of thought, (2) in its contemporary context, stylized thought constraint.*

Even unscientific statements contain compulsory connections.

Consider a myth, such as the Greek myth of Aphrodite, Hephaistos, and Ares. Aphrodite cannot but be the wife of Hephaistos and the lover of Ares. As any poet knows, a web of fantasy spun for long enough always produces inevitable, "spontaneous" substantive and formal connections. In a romance about chivalry, for instance, one cannot simply write "horse" instead of "steed," although these words are logically synonyms differing only in style. There are consequential links in musical imagination too, which correspond to the example: "Assuming $O = 16$ then $H = 1.008$." An artistic painting also exhibits its own constraining style. This we can easily demonstrate by placing part of a second painting over a good painting executed in a definite style. The two parts would clash with each other, even if the two paintings were matched in content. Thus every product of intellectual creation contains relations "which cannot exist in any other way." They correspond to the compulsory, passive links in scientific principles. These relations can be, as it were, objectivized and regarded as expressions of "beauty" or "truth." There actually are special individual and collective conditions which favor just such objectivization.

In the field of cognition, *the signal of resistance* opposing free, arbitrary thinking is called a *fact.* * This notice of resistance merits the adjective "thought collective," because every fact bears three different relations to a thought collective: (1) *Every fact must be in* *line with the intellectual interests of its thought collective,* since resistance is possible only where there is striving toward a goal. Facts in aesthetics or in jurisprudence are thus rarely facts for science. (2) *The resistance must be effective within the thought collective. It must be brought home to each member as both a thought constraint and a form to be directly experienced.* In cognition this appears as the connection between phenomena which can never be severed within the collective (see chap. 3 at note 26). This linkage seems to be truth and conditioned only by logic and content. Only an investigation in comparative epistemology, or a simple comparison after a change has occurred in the thought style, can make these inevitable connections accessible to scientific treatment. The principle of immutability of species characteristics was

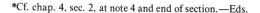

*Cf. chap. 4, sec. 2, at note 4 and end of section.—Eds.

valid for classical bacteriology, according to the interpretation of the time. If a scientist of that time had been asked why the principle was accepted or why the characteristics of species were conceived in this way, he could only have answered, "Because it is true." Only after a change in thought style did we learn that the opinion was constrained mainly by the methods applied. The passive linkage between these principles was transformed into an active one (cf. the definition in chap. 1, p. 8).[5] (3) *The fact must be expressed in the style of the thought collective.*

The fact thus defined as a "signal of resistance by the thought collective" contains the entire scale of possible kinds of ascertainment, from a child's cry of pain after he has bumped into something hard, to a sick person's hallucinations, to the complex system of science.

Facts are never completely independent of each other. They occur either as more or less connected mixtures of separate signals, or as a system of knowledge obeying its own laws. As a result, every fact reacts upon many others. Every change and every discovery has an effect on a terrain that is virtually limitless. It is characteristic of advanced knowledge, matured into a coherent system, that each new fact harmoniously—though ever so slightly—changes all earlier facts. Here every discovery is actually a re-creation of the whole world as construed by a thought collective.

A universally interconnected system of facts is thus formed, maintaining its balance through continuous interaction. This interwoven texture bestows solidity and tenacity upon the "world of facts" and creates a feeling both of fixed reality and of the independent existence of the universe. The less interconnected the system of knowledge, the more magical it appears and the less stable and more miracle-prone is its reality, always in accordance with the thought style of the collective.

The communal "carrier" of the thought style is designated the thought collective. The concept of the thought collective, as we use it to investigate the social conditioning of thinking, is not to be understood as a fixed group or social class. It is functional, as it were, rather than substantial, and may be compared to the concept of field of force in physics. A thought collective exists whenever two or more persons are actually exchanging thoughts. This type of thought collective is transient and accidental, forming and dissolving at any moment. But even this type induces a particular mood, which would otherwise affect none of the members and often recurs whenever these members meet again.

Besides such fortuitous and *transient* thought collectives there are *stable* or comparatively stable ones. These form particularly around organized social groups. If a large group exists long enough, the thought style becomes fixed and formal in structure. Practical performance then dominates over creative mood, which is reduced to a certain fixed level that is disciplined, uniform, and discreet. This is the situation in which contemporary science finds itself as a specific, thought-collective structure [*denkkollektives Gebilde*].

A thought community [*Denkgemeinschaft*] does not fully coincide with the official community. The thought collective of a religion comprises all true believers, whereas the official religious community includes all the formally accepted members, irrespective of their way of thinking. It is thus possible to belong to the thought collective of a religion without being formally accepted as a member of that congregation, and vice versa. The internal structure and organization of a thought collective also differs from the organization of a community in the official sense. The intellectual leadership and the circles that form around it do not coincide with the official hierarchy and organization.

A closer investigation of thought style and of the general social characteristics of thought collectives in their mutual relations can be made by concentrating upon stable thought collectives. Such stable (or comparatively stable) thought communities, like other organized communes [*Gemeinden*],* cultivate a certain exclusiveness both formally and in content. A thought commune becomes isolated formally, but also absolutely bonded together, through statutory and customary arrangements, sometimes a separate language, or at least special terminology. The ancient guilds, for instance, are examples of such special thought communes. But even more important is the restricted content of every thought

---

Gemeinde: often used for the smallest administrative district of local government in some European countries.—Eds.

collective as a special realm of thinking. There is an apprenticeship period for every trade, every religious community, every field of knowledge, during which a purely authoritarian suggestion of ideas takes place, irreplacable by a "generally rational" organization of ideas. The optimum system of a science, the ultimate organization of its principles, is completely incomprehensible to the novice. Yet this is the only valid yardstick for the expert. We have already described this situation in the case of the closure of thought within serology, which has only a traditional and not a "rational" initiation.

Every didactic introduction is therefore literally a "leading into" or a gentle constraint. The history of science is pedagogically helpful, because long-established concepts have the advantage of less thought specialization and are therefore more easily understood by the novice. Furthermore, the public at large, and therefore many an apprentice, are already familiar with them. The initiation into any thought style, which also includes the introduction to science, is epistemologically analogous to the initiations we know from ethnology and the history of civilization. Their effect is not merely formal. The Holy Ghost as it were descends upon the novice, who will now be able to see what has hitherto been invisible to him. Such is the result of the assimilation of a thought style.

The organic exclusiveness of every thought commune goes hand in hand with a stylized limitation upon the problems admitted. It is always necessary to ignore or reject many problems as trifling or meaningless. Modern science also distinguishes "real problems" from useless "bogus problems." This creates specialized valuation and characteristic intolerance, which are features shared by all exclusive communities.

Corresponding to any thought style is its practical effect or application. Any thought can be applied. Even the confirmation or refutation of conjectures calls for mental activity. Verification is therefore just as much bound by thought style as is assumption. Thought constraint, habits of thought, or, at least, a definite aversion to alien thinking that does not conform to a given thought style all help to guard the harmony between application and thought style. Guild associations are communities that are clearly directed to practical aims. It is instructive to see how differently, depending on the nature of the trade, similar practical problems are solved. A crack in the wall plaster, for instance, presents a painter with a problem different from that which a bricklayer has to face. The painter sees only the surface damage and treats it accordingly. But the bricklayer worries about the wall structure and is likely to "work in depth." The way in which their thinking is stylized is revealed by the way it is applied.[6]

Independently of the possible organization in form and content of a stable collective, such as has been noted for the organization of a church community or a trade union, there are also structural characteristics shared by all such communities of thought. The general structure of a thought collective consists of both a small esoteric circle and a larger exoteric circle, each consisting of members belonging to the thought collective and forming around any work of the mind [*Denkgebilde*], such as a dogma of faith, a scientific idea, or an artistic musing. A thought collective consists of many such intersecting circles. Any individual may belong to several exoteric circles but probably only to a few, if any, esoteric circles. There is a graduated hierarchy of initiates, and many threads connecting the various grades as well as the various circles. No direct relation exists between the exoteric circle and that creation of thought [*Denkgebilde*] but only one mediated esoterically. Thus most of the members of the thought collective are related to the works produced by the thought style [*Gebilde des Denkstiles*] only through trusting the initiated. But the initiated are by no means independent. They are more or less dependent, whether consciously or subconsciously, upon "public opinion," that is, upon the opinion of the exoteric circle. This is generally how the intrinsic self-containment of the thought style with its inherent tenacity arises.

The esoteric circles thus each enter into a relation with their exoteric circles known in sociology as the relation of the elite to the masses. If the masses occupy a stronger position, a democratic tendency will be impressed upon this relation. The elite panders, as it were, to public opinion and strives to preserve the confidence of the masses. This is the situation in which the thought collective of science usually finds itself today. If the elite enjoys the stronger position, it will endeavor to maintain distance and to isolate itself from the crowd. Then secretiveness and dogmatism dominate the life of the thought collective. This is the situation of religious thought collectives. The first, or democratic, form must lead to the development of ideas and to progress, the second possibly to conservatism and rigidity.

Individuals too take up special mutual positions in the communication of thoughts within a collective. If there exists a relation of definite mental superordination and subordination between two individuals, as between teacher and pupil, it is really not a relation between individuals but between elite and masses. On the one hand there is basically trust, and on the other, dependence on public opinion and "commonsense." Between two members of the same thought collective on the same mental level, there is always a certain solidarity of thought in the service of a superindividual idea which causes both intellectual interdependence and a shared mood between the two individuals. No question, once raised, can remain totally without effect. Each is pondered and has a place within the thought style. This comradeship of mood can be sensed after only a few sentences have been uttered and makes true communication possible. Without it, the speakers are at cross purposes. A special feeling of dependence therefore dominates all communication of thought within a collective. The general structure of a thought collective entails that *the communication of thoughts within a collective, irrespective of content or logical justification, should lead for sociological reasons to the corroboration of the thought structure [Denkgebilde].* * Trust in the initiated, their dependence upon public opinion, intellectual solidarity between equals in the service of the same idea, are parallel social forces which create a special shared mood and, to an ever-increasing extent, impart solidity and conformity of style to these thought structures [Denkgebilde].** The greater the distance in time or space from the esoteric circle, the longer a thought has been conveyed *within the same thought collective,* the more certain it appears. If

---

*In this context the ambiguous "thought structure" is selected, since an indirect reference to the patterns of thought *may* be implicit in this direct reference to the products of thought.—Eds.
**Thought products and the thought style under which these arise are both of them socially constrained. Cf. Preface.—Eds.

the bonds consist in mental training during childhood years or, better still, in a tradition several generations old, they will be indissoluble.

At a certain stage of development the habits and standards of thought will be felt to be the natural and the only possible ones. No further thinking about them is even possible. But once they have entered personal consciousness, they can also be regarded as supernatural, a dogma, a system of axioms, or even a useful convention. In this context it would be of interest to compare the history of science or the history of sports from semireligious practices in antiquity to the health-oriented sports of our own day.

The complex structure of modern society results in multiple intersections and interrelations among thought collectives both in space and time. We see professional and semiprofessional thought communities in commerce, the military, sports, art, politics, fashion, science, and religion. The more specialized a thought community is and the more restricted in its content, the stronger will be the particular thought nexus among the members. It breaks down boundaries of nation and state, of class and age. Compare the social role of sports or of spiritualism. Special terms such as *match, foul,* and *walkover* in sports; *demarche* and *exposé* in politics; *Saldo* [balance], *Konto* [account], *hausse* [bulls], and *baisse* [bears] on the Stock Exchange; *staffage* [props] and *expression* in the arts, each within its own thought collective, are used even across the barriers of national languages. The printed word, film, and radio all allow the exchange of ideas within a thought community. They also make possible the connection between the esoteric and the exoteric circles even across long distances and in spite of little personal contact.

A good example of the general structure of a thought collective is provided by the thought community of the world of fashion, as long as we examine only the common mental outlook of the followers of fashion and disregard either the general economic and social factors or the special professional and commercial factors of that field. What is of interest is fashion consciousness as such, independent of the content of fashion. The special mood of the thought collective of fashion is constituted by a readiness immediately to notice that which is fashionable and to consider it of absolute importance, by a feeling of solidarity with other members of the collective, and by an unbounded confidence in the members of the esoteric circle. The most dedicated followers of fashion are found far out in the exoteric circle. They have no immediate contact with the powerful dictators forming the esoteric circle. Specialized "creations" reach them only through what might be called the official channels of intracollective communication, depersonalized and thus all the more compulsive. Nothing is motivated in petty style; they are simply told "ce qu'il vous faut pour cet hiver" [what you need for this winter], or "à Paris la femme porte" [in Paris, women are wearing], or "Lancé au printemps par quelques jeunes femmes de la société parisienne" [presented to the public in the spring by several young ladies of Parisian society]. It is coercion of the strongest kind, because it appears in the guise of a self-evident necessity and is thus not even recognized as a coercive force. And woe to the true believer who does not or cannot conform. She feels cast out and branded, because she knows full well that every fellow member of the collective immediately notices her act of treason. For the esoteric members the coercion is much reduced. They can permit themselves many a new-fangled idea, which does not become a "must" until subsequent communication has taken place throughout the thought collective. But they too are held by the style of their own creations to particular "obligatory matchings": baroque sleeves may not be worn with an Empire waistline, to name only one example.

If we compare various thought styles, we can easily see that the differences between two such thought styles can be greater or smaller. The thought style of the physicists, for instance, does not differ all that much from that of the biologists, unless the latter happen to adhere to the thought style of the vitalists. There is a much greater difference in style between the physicists and the philologists, and a much greater one still between a modern European physicist and a Chinese physician or a cabalistic mystic. Here the divergence between thought styles is so wide that in comparison, the divergence between the thought styles of the physicist and of the biologist dwindles into nothing. One could actually speak of nuances of style, of varieties in style, and of different styles. But it is not the aim of this book to construct a complete theory of thought styles. All I want to do is point out a few distinctive properties of the communication of thoughts between collectives.

The greater the difference between two thought styles, the more inhibited will be the communication of ideas. Collectives, if real communication exists between them, will exhibit shared traits independent of the uniqueness of any particular collective. The principles of an alien collective are, if noticed at all, felt to be arbitrary and their possible legitimacy as begging the question. The alien way of thought seems like mysticism. The questions it rejects will often be regarded as the most important ones, its explanations as proving nothing or as missing the point, its problems as often unimportant or meaningless trivialities. Depending upon the relation between the collectives, single facts and concepts are considered either free inventions, which scientists simply ignore like, for instance, "psychic facts" [*spiritistische Tatsachen*]. Less divergent collectives, alternatively, may produce only different interpretations, translations into another dialect of thought, as, for instance, theologians would translate these same psychic facts. Scientists have similarly adopted many individual alchemic facts. So-called commonsense, as the personification of the thought collective of everyday life, has become in this same way a universal benefactor for many specific thought collectives.

Words as such constitute a special medium of intercollective communication. Since all words bear a more or less distinctive coloring conforming to a given thought style, a character which changes during their passage from one collective to the next, they always undergo a certain change in their meaning as they circulate intercollectively. One could compare the meaning of the words "force," "energy," or "experiment" for a physicist, a philologist, or a sportsman; the word "explain" for a philosopher and a chemist, "ray" for an artist and a physicist, or "law" for a jurist and a scientist.

In summary, the intercollective communication of ideas always results in a shift or a change in the currency of thought. Just as the shared mood within a thought collective leads to an enhancement of thought currency, so does the change in mood during the intercollective passage of ideas produce an adjustment in this cash value across the entire range of possibilities, from a minor change in coloration, through an almost complete change of meaning, to the destruction of all sense. Compare the fate of the philosophical term "absolute" in the thought collective of scientists.

In chapter 1 we described the passage of the syphilis concept from one thought community to another. Each passage involved a metamorphosis and a harmonious change of the entire thought style of the new collective arising from the connection with its concepts. This change in thought style, that is, change in readiness for directed perception, offers new possibilities for discovery and creates new facts. This is the most important epistemological significance of the intercollective communication of thoughts.

Something remains to be said about the individual's belonging to several thought communities and acting as a vehicle for the intercollective communication of thought. The stylized uniformity of his thinking as a social phenomenon is far more powerful than the logical construction of his thinking. Logically contradictory elements of individual thought do not even reach the stage of psychological contradiction, because they are separated from each other. Certain connections, for instance, are considered matters of faith and others of knowledge. Neither field influences the other, although logically not even such a separation can be justified. A person participates more often in several very divergent thought collectives than in several closely related ones. There were and still are physicists, for instance, who profess the religious or spiritualist thought style, but few of them have been interested in biology once it became an independent discipline. Many physicians are engaged in historical or aesthetic studies but only a few in natural science. If thought styles are very different, their isolation can be preserved even in one and the same person. But if they are related, such isolation is difficult. The conflict between closely allied thought styles makes their coexistence within the individual impossible and sentences the person involved either to lack of productivity or to the creation of a special style on the borderline of the field. This incompatibility between allied thought styles within an individual has nothing to do with the delineation of the problems toward which such thinking is directed. Very different thought styles are used for one and the same problem more often than are very closely related ones. It happens more frequently that a physician simultaneously pursues studies of a disease from a clinical-medical or bacteriological viewpoint together with that of the history of civilization, than from a clinical-medical or bacteriological one together with a purely chemical one.

As I select out of an abundance of data these few phenomena concerning the communication of ideas, I am fully aware of the fragmentary nature of my presentation. But they may suffice to demonstrate to science-oriented theoreticians, in particular, that even the simple communication of an item of knowledge can by no means be compared with the translocation of a rigid body in Euclidean space. Communication never occurs without a transformation, and indeed always involves a stylized remodeling, which intracollectively achieves corroboration and which intercollectively yields fundamental alteration. Those who fail to grasp this point will never reach a positive epistemology.[7]

4. Some Characteristics of the Thought Collective of Modern Science

In the previous section we described the general structure of thought collectives—their esoteric and exoteric circles, and the general rules of intra- and intercollective communication of thought. We shall now discuss the special structure of the thought collective of modern science, particularly the effect of both the esoteric circle and the exoteric circle within the framework of science. We shall disregard characteristic features of any specialized thought collective such as that of the physicists or that of the sociologists, because the structure of modern Western science has many common features.

Take the case of a researcher who creatively approaches a problem and is a "specialized expert" informed in the greatest depth—for example, a radium specialist in the science of radioactivity. He constitutes the center of the esoteric circle of this problem. The circle includes, as "general experts," scientists working on related problems—all physicists, for instance. The exoteric circle comprises the more or less "educated amateurs." A contrast between *expert* and *popular* knowledge is hence the first effect of the general structure of the thought collective in science. The richness of this field requires that even within the specialized esoteric circle, a sphere of special experts must be distinguished from that of general ones. Let us then consider both *journal science* and *vademecum science,* which together constitute expert science. Because initiation into science is based on special methods of teaching, we must list *textbook science* as yet a fourth socio-intellectual form, which, however, is less important in our context.

Let us begin the discussion of these circles by considering *popular science.* This furnishes the major portion of every person's knowledge. Even the most specialized expert owes to it many concepts, many comparisons, and even his general viewpoint. It thus constitutes the general operative factor in cognition and must accordingly rank as an epistemological problem. When an economist speaks of the *organism* of the economy, or a philosopher of *substance,* or a biologist of the *syncytium* [*Zellstaat,* lit. "cell state"], they use, each within his own discipline, concepts derived from their fund of popular knowledge. They build up their specialized sciences around these concepts. We shall presently have the opportunity of repeatedly finding items of popular knowledge from other fields within the depths of these sciences. Such items have often set the standard for the content of expert knowledge and have determined its development for decades.

Popular science is a special, complex structure. Since speculative epistemologists have never investigated actual knowledge but only a fanciful picture of it, an epistemological investigation of popular science has yet to occur, at least so far as I am aware. But this is not the place to close this gap; a few hints must suffice.

Popular science in the strict sense is science for nonexperts, that is, for the large circle of adult, generally educated amateurs. It cannot therefore be classed as introductory science. Normally a textbook, not a popular book, is used for purposes of introduction. Characteristic of the popular presentation is the omission both of detail and especially of controversial opinions; this produces an artificial simplification. Here is an artistically attractive, lively, and readable exposition with last, but not least, the apodictic valuation simply to accept or reject a certain point of view. Simplified, lucid, and apodictic science—these are the most important characteristics of exoteric knowledge. *In place of the specific constraint of thought by any proof, which can be found only with great effort, a vivid picture is created through simplification and valuation.* The ultimate aim of popular knowledge is a Weltanschauung: a special structure [*Gebilde*] emerging from an emotive selection of popular knowledge from various fields.

Little as any Weltanschauung can meet the demands of specialized knowledge, it does provide the background that determines the general traits of the thought style of an expert. This may sometimes be no more than an exalted feeling about the solidarity of all human knowledge. Or it may be a belief either in the possibility of a universal science or in the albeit limited potential for further development in science. This closes the circle of intracollective dependence in knowledge. Popular exoteric knowledge stems from specialized esoteric knowledge. Owing to simplification, vividness, and absolute certainty it appears secure, more rounded, and more firmly joined together. It shapes specific public opinion as well as the Weltanschauung and in this form reacts in turn upon the expert.

A good example of this situation is provided by a bacteriological examination, where the findings are recorded in the diagnostic laboratory by the esoteric expert team for the exoteric general practitioner. The diagnosis of a specimen from a throat swab, for instance, reads as follows: "The microscopic specimen shows numerous small rods whose shapes and positions correspond to those of diphtheria bacilli. Cultures grown from them produced typical Löffler bacilli." This finding is specially written to suit the general practitioner, but it does not represent the knowledge of the expert. It is vivid, simplified, and apodictic. The general practitioner can rely upon it. But the expert reporting *the same finding* to another expert would write in the following terms. "*Microscopic aspect:* numerous bacilli, many of which are club-shaped and slightly curved, others rather slim and straight or uncharacteristically plump. Arrangement in several places finger- and pallisade-shaped, elsewhere singular and irregular. Gram-positive. Several bacilli Neisser-positive. Löffler methyl blue: many lacerated bacilli. *Culture:* Costa medium: purplish-red, slightly smeary, sharply defined colonies, in which bacilli were found mostly typical in staining characteristics, morphology, and arrangement. *Toxin production* and neutralization tests were not performed. In view of the origin of the examined material, and the morphological and culture characteristics of the bacilli, the diagnosis of Löffler bacilli seems sufficiently well established."

This version, although theoretically far more precise, would not appeal to a general practitioner, least of all the passage according to which the origin of the examined material is considered one of the supports (and, indeed, one of the important supports) for the conclusion. "What is going on? I just asked you what this throat swab really contains and you reply: because it is a throat swab the conclusion is justified that it is diphtheria. That is being mischievous. I wanted your support, but you went and used me to support yourself." Yet this expert finding is already purposively simplified and apodictic in many respects. Everything that is unimportant from a scientific viewpoint is omitted, such as accompanying bacteria—or what are currently thought to be unimportant accompanying bacteria. The vagueness of the limits of speciation for corynebacteria also remains unconsidered. The conclusion that the rods found in the microscopic specimen of the swab are identical with those in the culture is actually a complicated, specialist thought construction, although it is presented here as just a simple fact. Furthermore, the case is extremely elementary. It is not very often that everything works in such perfect agreement. Frequently the arrangement of the bacilli is not quite so typical. Staining is not always so unambiguous, for it can be positive, negative, or indeterminate. Finally, the culture may contradict the microscopic specimen.

No matter how a given case may be described, the description is always a simplification permeated with apodictic and graphic elements. *Every communication and, indeed, all nomenclature tends to make any item of knowledge more exoteric and popular.* Otherwise each word would require a footnote to assign limitations and provide explanations. Each word of the footnote would need in turn a second word pyramid. If continued, this would produce a structure that could be presented only in multidimensional space. Such exhaustive expert knowledge completely lacks clarity and is unsuitable in any practical case. It must be remembered that such a pyramidal structure does not yield more general and recurrent elements, which would basically simplify construction if they could be described separately. We always remain within the same stratum of concepts, equidistant from "fundamental concepts," the possible construction of which constitutes a cognitive effort in its own right and presents the same difficulties. *Certainty, simplicity, vividness originate in popular knowledge.* That is where the expert obtains his faith in this triad as the ideal of knowledge. Therein lies the general epistemological significance of popular science.

Our example presents a part of exoteric science which is still very close to the esoteric center. The general practitioner is not all that far removed from the bacteriological specialist. If we proceed to the large circle of the "generally educated," knowledge becomes pellucid and facile; at the same time, thought-constrained proofs disappear: it becomes even more apodictic. The mother of the child whose throat swab had been examined is simply informed: "Your child has diphtheria."

The following popular description of the classical period of bacteriology is found in Gottstein's excellent book on epidemiology.[8] "An examination was carried out on a patient, or on a susceptible animal made ill through inoculation with products of the disease in question. Certain micromycetes were found here which were proved absent in other diseases. Methods were devised for obtaining a pure culture of them on suitable artificial media. Many generations of the germ were grown on this culture medium with the strictest prevention of any contamination by other schizomycetes. Their properties were studied and the disease was reproduced by inoculating other animals. The chain of proof was thereby completed. Production of the characteristic disease has always been successful in isolated experiment and remains so today." How simple, certain, and lucid does this bacteriological discovery appear! The description can certainly not be replaced by a better *popular* version. As a "general scheme" it is basically correct. It just does not correspond to detailed expert knowledge. Apart from ignoring the many restrictions and complications as well as the contradictory views and errors of the research workers, this presentation completely conceals the interaction between the genesis of a discovery and the genesis of concepts. The description reads as if definitive concepts and ideas existed a priori. The concept of a disease entity of "*certain* micromycetes" is an example, as are the pure culture and the connection between disease and microbes. It is as if their "consistent" application alone led to the discovery and no other concepts were possible. Truth is thus made into an objectively existing quality. Scientists are accordingly divided into two classes; the "bad guys," who miss the truth, and the "good guys," who find it. This valuation, which is a general trait of exoteric thinking, was also created by the demands of the intracollective communication of thought and subsequently reacts upon expert knowledge.

Let us introduce another example. On page 5 of Gottstein's book, the history of syphilis is described as follows.

In 1495 a disease erupted suddenly and with unprecedented violence, spreading among the French mercenaries fighting in Italy, who quickly carried this "syphilis" across the whole of Europe. The rapid spread of this epidemic soon suggested that a new disease was involved. The suspicion naturally arose that it had been introduced from newly discovered America, where it was known at the time to exist, although in a milder form. Controversy still rages today over the American provenance of syphilis. Alternatively it is claimed that syphilis was already found in the Old World in antiquity. Be that as it may, at the end of the fifteenth century it spread unusually widely and with great severity. From that time to our own day, syphilis has never lost its significance as a common disease, although its manifestations have changed greatly.

How simplified and crystal-clear is this history! Where is the assiduous elaboration of the specific disease concept of "syphilis"? The whole metamorphosis of the thought style from the fifteenth to the twentieth century and the historico-cogitative as well as sociocogitative dependence of its stages have become invisible in the description. From descriptions such as this, the general conviction follows that there is no development of thought. This is a conviction that in turn also influences the expert, and it is decisive for the epistemologists who regard their task exclusively as the treatment of the question of "right" or "wrong" knowledge.

The achievement of vividness in any knowledge [*eines Wissens*] has a special inherent effect. A pictorial quality is introduced by an expert who wants to render an idea intelligible to others or for mnemonic reasons. But what was initially a means to an end acquires the significance of a cognitive end. The image prevails over the specific proofs and often returns to the expert in this new role. We can study this phenomenon well by looking at the effect of Ehrlich's clear symbolism mentioned in chapter 3. The lock-and-key symbols became the theory of specificity and for a long time dominated the very depths of the specialized science of serology.

In addition to such general influences fed back from popular science, every field contains many specific influences. As an example, the whole lipoid theory of the Wassermann reaction is founded on a popular chemical concept of the lipoid bodies, which is in no way identical with the specialized chemical one. We thus have the strange phenomenon that the lipoid of the serologist differs from that of the chemist, just as the concept of "state" in biology, which construes the organism as a syncytium, a "cell state," is very different from the state as construed by political science.

If we move still further away from the esoteric center toward the exoteric periphery, thinking appears to be even more strongly dominated by an emotive vividness that imparts to knowledge the subjective certainty of something holy or self-evident. No more thought-constraining proofs are demanded, for the "word" has already become "flesh." As an example of such grossly popular science, consider an illustration representing the hygienic fact of droplet infection. A man emaciated to a skeleton and with greyish-purple face is sitting on a chair and coughing. With one hand he is supporting himself wearily on the arm of the chair, with the other he presses his aching chest. The evil bacilli in the shape of little devils are flying from his open mouth. . . . An unsuspecting rosy-cheeked child is standing next to him. One devil bacillus is very, very close to the child's mouth. . . . The devil has been represented bodily in this illustration half symbolically and half as a matter of belief. But he also haunts the scientific speciality to its very depths, in the conceptions of immunological theory with its images of bacterial attack and defense.

In contrast with popular science, whose aim is *vividness,* professional science in its *vademecum* (or handbook) form requires a *critical synopsis in an organized system.*

In the history of the discovery of the Wassermann reaction and in the chapter on observation and experiment, we have attempted to describe the creative expert as the personified intersection of various thought collectives as well as of various lines of development of ideas and as a personal center of new ideas. The report that he writes has, in the first instance, a form we may call *journal science.*

Any attempts to organize journal science into a unified whole will soon encounter difficulties. The various points of view and working methods are so personal that no organic whole can be formed from the contradictory and incongruent fragments. It is not possible to produce a vademecum simply from a collection of articles that have appeared in journals. Only through the socio-cognitive migration of fragments of personal knowledge within the esoteric circle, combined with feedback [*Rückwirkung*] from the exoteric circle, are these fragments altered so that additive, impersonal parts can arise from the nonadditive personal ones.

Journal science thus bears the imprint of the provisional and the personal. Its first feature is that despite a pronounced limitation of the problems under consideration there is always an urge to link up with the entire complex of problems associated with the field in question. Any paper published in a scientific journal contains in the introduction or conclusion just such a connection with vademecum science as proof that the author aims at incorporating his paper in the vademecum and regards its present state as provisional. This provisional aspect also comes through in the details about both plans and hopes as well as in the polemics. There is also a specific caution characteristic of journal contributions. This can be recognized in the typical turns of phrase such as, "I have *tried* to prove...," "*It appears* possible that...," or negatively, "It was not possible to prove that..." Such jargon serves to shift the "holy of holies" of science, that is, any judgment about the existence or non-existence of a phenomenon, from the individual worker to the solely authorized collective. Only in impersonal vademecum science will we find expressions such as, "This exists or that does not," or "This or that exists," or "It has been firmly established that ..."

It is as if every competent scientist required, in addition to the control inherent in the style conformity of his work, a further control and processing by the collective. It is as if he were aware that only intracollective communication of thought can lead from cautious uncertainty to certainty.

The personal aspect is a second feature of journal science somehow related to the first. The fragmentary nature of the problems, the contingency of the material (as, for instance, casuistics in medicine), the technical details, in short, the uniqueness and novelty of the working material tend to associate it inseparably with the author. Every researcher is aware of this but at the same time feels that any such personal element in the work is a fault. He almost always wants to hide his identity. This is recognizable, for instance, in the characteristic "we" instead of "I," the specific "plural of modesty," which is a hidden invocation of the collective. The specific discretion and duty of the individual research worker to remain in the background is formed from such modesty as well as from the characteristic caution just discussed.

Describable in terms of laboriously established, disjointed signals of resistance in thinking, this provisional, uncertain, and personally colored nonadditive journal science, then, is converted next into vademecum science by the migration of ideas throughout the collective. As we have already pointed out, this striving toward community, which expresses the dominance of the rank-and-file members of the thought collective of science over its elite, will be found in every work of the scientist. A "general verifiability" is officially demanded as a demagogic postulate, as it were. Yet, first it is not a general[9] examination but one by the thought collective, and, second, it consists solely in scrutinizing the stylization of any item of knowledge.

The vademecum is therefore not simply the result of either a compilation or a collection of various journal contributions. The former is impossible because such papers often contradict each other. The latter does not yield a closed system, which is the goal of vademecum science. A vademecum is built up from individual contributions through selection and orderly arrangement like a mosaic from many colored stones. The plan according to which selection and arrangement are made will then provide the guidelines for future research. It governs the decision on what counts as a basic concept, what methods should be accepted, which research directions appear most promising, which scientists should be selected for prominent positions and which should simply be consigned to oblivion. Such a plan originates through esoteric communication of thought—during discussion among the experts, through mutual agreement and mutual misunderstanding, through mutual concessions and mutual incitement to obstinacy. When two ideas conflict with each other, all the forces of demagogy are activated. And it is almost always a third idea that emerges triumphant: one woven from exoteric, alien-collective, and controversial strands.

In the history of the Wassermann reaction we described the process by which personal and provisional journal science becomes transformed into collective, generally valid vademecum science. This appears initially both as change of conceptual meaning and as reformulation of the problem, and subsequently as an accumulation of collective experience, the formation of a special readiness for directed perception and specialized assimilation of what has been perceived. Some of this esoteric communication of thought occurs already within the scientist himself. He conducts a dialogue with himself as he ponders, compares, and makes decisions. The less his decision rests on adaptation to vademecum science and the more original and bolder his personal thought style, the longer it will take to complete the process of collectivizing his results.

The following event may serve as an example of esoteric communication of thought within a transient collective. At a meeting of medical historians, the members discussed a case history contained in an ancient text and considered the possibility of making a modern diagnosis according to this old description. One of the members claimed that it would be impossible in the present case, because the methods of examination given by the author diverged too far from current ones. A second member argued that basically diagnosis was always possible, since after all the diseases themselves remained unchanged. All one had to do was to construct a picture of it from textual analysis. The first member rebutted by granting that the diseases certainly remained unchanged; but our training is different, and we simply cannot form a picture from so many emotive words that describe the gravity and horror of the disease but provide no objective clues for a diagnosis. It is true that the many terms in the text describe with extraordinary accuracy the patient's odor, the stratification of his excretions, changes in his perspiration, and even his cries of anguish. But we cannot even find out whether or not there was a fever. A lively discussion ensued lasting more than an hour, shifting from the casuistic to the fundamental. Strangely enough, however, the basic claim was maintained throughout that diseases as such, that is, disease entities, had not changed. This claim was a kind of lapse by the second ·speaker, and he admitted the point to me afterward. It became reinforced by the rather offhand confirmation by the first speaker and thereby acquired, oddly, the value of an axiom. But when the thought collective dissolved, not a single member of the discussion was prepared to take responsibility for it. The claim is doubtless untenable and was therefore only short-lived. But the rather impersonal mechanism of its origin, without anybody's deliberate intention or responsibility, can serve as a paradigm for typical principles of vademecum science. Very often it is impossible to find any originator for an idea generated during discussion and critique. Its meaning changes repeatedly; it is adapted and becomes common property. Accordingly it achieves a superindividual value, and becomes an axiom, a guideline for thinking.

A statement appears ipso facto more certain and more soundly established in the organized system of a discipline as presented in a vademecum than it does in any fragmentary description found in a journal. It becomes a definite thought constraint.

Let me give an example. The etiological concept of disease entity is not derived directly from individual contributions to the journals. Emerging originally from exoteric or popular ideas and from ideas formed outside the collective in question, it gradually acquired its present significance in the esoteric communication of thought and now forms one of the basic concepts of vademecum bacteriology. It could be attained only through a directed selection of individual investigations and a directed compilation. But once part of the vademecum, it is taught and generally used. It forms the keystone of the system and thus exerts a constraint on thinking. A statement such as the following becomes meaningless: "The

French pox, or syphilis, or the carnal scourge, which is the consequence of contagious and leprous affections of the genitals, is a
daughter of leprosy and can in certain circumstances in turn become the mother of leprosy."[10] Yet it is meaningless only for our
thought style. According to the etiological concept of disease,
syphilis is a spirochaetosis and leprosy a disease caused by the
specific bacillus, so that no relation exists between the two diseases. If, however, the diseases are defined symptomatically, their
relation becomes undeniable and the statement deeply meaningful.
It has been explained that the etiological concept of disease is not the
only logically possible one. Nor does it just arise spontaneously in
the presence of a certain quantity of knowledge. Nevertheless contemporary scientists, or most of them, are constrained by this
concept and cannot think in any different way. This also affects the
whole of pathology and bacteriology. The latter has become a
medical science and has almost severed its connection with botany.
The thought style of pathology in general and of bacteriology is
therefore nonbiological, a point that manifests itself both in methodology[11] and in the narrowness of the problem complex with its
strict limitation to medical applications.

The situation is very similar as regards the formulation of the
modern concept of the chemical element, which is based on proportions by weight. This concept is also the result of truly collective
work, which began with esoteric communication of thought about
individual contributions. It thus became systematic and impersonal vademecum science. "After Boyle's day, however, it gradually came to be seen that certain substances resisted all such
attempts to change them into others without increasing their
weight. For example, all changes which iron can be made to
undergo are accompanied by an increase in the weight of the
iron. . . . It was slowly discovered that at least seventy such different substances must be classed as elements."[12] Lavoisier contributed a great deal to this concept of element. It was actually
during his lifetime that proportions by weight came to be accepted
as stable relations. In describing these events, Ostwald mentioned a
"strange psychological phenomenon, which occurs very often at
moments of important progress in science."[13] It was indeed
Lavoisier who, with his theory of combustion and his law of the

conservation of weight, provided the necesssary support for the idea that proportions by weight were decisive in formulating the concept of element. Yet it was this same Lavoisier who introduced such imponderable elements as heat and light in addition to the ponderable elements and who thus "contradicted his own idea." Ostwald, maintaining a completely individualistic psychological point of view, could explain this strange phenomenon only in terms of psychology. He stated that often "the ultimate step, which confirms a new idea and rejects old ones, is precisely the one which remains unnoticed and neglected by the creator of the new idea." He tries to account for this in terms of the exhaustion of the researcher, who has no strength left for this last refinement of his idea. I believe that our observations so far have shown clearly that this incongruence between an idea as examined retrospectively and the description given by the "originator" himself, that is, by the research worker concerned, can be explained simply by the fact that the true creator of a new idea is not an individual but the thought collective. As has been repeatedly stressed, the collective remodeling of an idea has the effect that, after the change in thought style, the earlier problem is no longer completely comprehensible. That the modern concept of the chemical element also has a prehistory is well known. Like that of the etiological concept of disease it can be traced back to the mythical age. In this case again, the modern vademecum version is derived from alien collectives, exoteric sources, and esoteric communication of thought. These examples, to which similar ones could be added at will, make the role of vademecum science obvious. This is the means by which exoteric knowledge, knowledge originating in other collectives, and strictly specialist knowledge are all selected, blended, adapted, and then molded into a system. Concepts originating in this manner become dominant and binding on every expert. The preliminary signal of resistance has become thought constraint, which determines what cannot be thought in any other way, what is to be neglected or ignored, and where, inversely, redoubled effort of investigation is required. The readiness for directed perception becomes consolidated and assumes a definite form.

This relation between journal science and vademecum science shows up in modern progressive science as a characteristic structure of the esoteric circle. It resembles a column of troops on the march. Every discipline, in fact almost every problem, has its own *vanguard,* the group of research scientists working practically on a given problem. This is followed by the *main body,* the official community. Then come the somewhat disorganized stragglers. This structure becomes the more conspicuous the greater the progress in the field of investigation. Journal science, which comprises the latest work, becomes more or less removed from vademecum science, which always lags behind. The vanguard does not occupy a fixed position. It changes its quarters from day to day and even from hour to hour. The main body advances more slowly, changing its stand—often spasmodically—only after years or even decades. Its path does not closely follow that of any one of the vanguards. The main body adjusts its advance according to reports received from the vanguard, but maintains a certain independence. The direction that the main body actually chooses from the many suggested by the vanguards is always unpredictable. Paths must first be widened into roads, and the ground leveled, so that the terrain undergoes considerable change before it can become the garrison of the main body.

This indubitable phenomenon is obviously social in character, and has important theoretical consequences. If a scientist is asked about the status of a given problem, he must first specify the vademecum view as something impersonal and comparatively fixed, although he knows full well that it is inevitably already out of date. The various views of his fellow researchers currently working on the problem must be added to this, but only as their *personal* views, even though he knows that some of these may one day form the future vademecum views. It is characteristic of the social nature of science that it takes a typically binding position (sometimes this is less typical and more provisional) on almost any problem. It is particularly important, epistemologically, that the binding position is considered more exoterically conditioned than the provisional one. This indicates the dominance of the mass over the elite in a democratic thought collective.

If a fact is taken to mean something fixed and proven, it exists only in vademecum science. The preliminary stage of disjointed signals of resistance within journal science really constitutes only the predisposition for a fact. Later, at the stage of everyday popular knowledge, the fact becomes incarnated as an immediately perceptible object of reality.

5. Thought Styles

The special *thought style* of the thought collective of modern science becomes intelligible against this specific structural background. To make the concept of thought style even more tangible and more familiar it is appropriate to compare the modern scientific version with a few older ones.

In those days, to quote the words of Dr. Samuel Brown, the metals were suns and moons, kings and queens, red bridegrooms and lily brides. Gold was Apollo, sun of the lofty dome; silver, Diana, the fair moon of his unresting career, and chased him meekly through the celestial grove; quicksilver was the wing-footed Mercury, Herald of the Gods, new-lighted on a heaven-kissing hill; iron was the ruddy-eyed Mars, in panoply complete; lead was heavy-lidded Saturn, quiet as a stone, within the tangled forest of material forms; tin was the Diabolus Metallorum, a very devil among the metals, and so forth in not unmeaning mystery.—There were flying birds, green dragons, and red lions. There were virginal fountains, royal baths, and waters of life. There were salts of wisdom, and essential spirits . . . , etc.[14]

This is how chemistry was described before it entered the modern age. Such mystical allegories and comparisons and the strongly emotional images exhale an atmosphere that is completely alien to our scientific thinking. The comparison of gold with the sun and of silver with the moon survives only in popular imagination. Associating lead with Saturn and tin with the devil has lost all meaning even in popular thinking. It is a special, self-contained style, consistent from its own point of view. Those people thought and saw differently than we do. They accepted certain symbols that to us appear fanciful and contrived. What if we could present our symbols—the potential, or physical constants, or the gene of heredity, etc.—to thinkers of the Middle Ages? Could we expect them to be delighted with the "correctness" of these symbols and instantly listen to reason? Or, conversely, would they find our symbolism just as fanciful, contrived, and arbitrarily devised as we find theirs?

If we want to investigate an earlier thought style, we must examine the original sources, not modern summaries of old viewpoints. Consider a passage from Paracelsus:[15] "If you have faith small as a mustard-seed and you are yet earthly spirits, how much higher would you be if your faith were large as melons. Again, how far should we surpass the spirits, if our faith were like huge pumpkins." To illustrate the strength or weakness of faith by comparing it with mustard-seed can be accepted, if only because of biblical tradition,* so long as we remain conscious of its metaphorical character. But that it should be possible to establish a scale or a system by which to measure the strength of a person's faith against objects of various sizes is an idea we find startling. Anyone could use this sentence, for instance: "It is bad if you refuse to deviate from your demands by a finger's breadth." But the following sentence appears impossible to use in a sober frame of mind: "It is bad if you refuse to deviate from your demands by even a finger's breadth, when it is actually necessary that you should deviate from them by a foot or even a yard." For to us this sentence is either eccentric poetry or a foolish fancy of using geometrical yardsticks for psychological events. And what did Paracelsus do? Did he consider his faith-measuring system only a metaphor, or an adequate measuring system as well? This becomes clear in his treatise on the begetting of sensitive things in reason.[16] "As long as the womb has a seed within it, it no longer draws another into it. It must only remain quiet and consummate, and it will be fertile. But when it becomes cold in old age, nothing more will happen, once the drawing power dies in the cold." He explains the infertility of old women in terms of the coldness of old age, which makes the (apparently temperature-sensitive) seed-drawing power of the womb die. Coldness of old age is to him not a metaphorical circumlocution for frigidity, but absolutely identical with physical cold. We often also read in ancient writings that ravenous hunger [*Heisshunger*, literally "hot-hunger"] cooks raw food as fire does and thus makes it digestible.

In a book published two hundred years later we read:[17] "Why is

*Matthew 17:19–20.—Eds.

a person with an empty stomach heavier than after a meal? Because eating increases the quantity of the spirits, which owing to
their airy and fiery nature lighten the human body; for fire and air
generally do this. For the same reason a cheerful person is much
lighter than a sad one, because a cheerful one harbors more of
these little spirits than a sad one. A dead person is also much
heavier than a living person, because the latter is full of little spirits
whereas the former is deprived of them." The feeling of heaviness
(sluggishness) as well as the modern concept of physical weight,
heavy-heartedness, and even the difficulty (unwieldiness) of lifting
a corpse are here regarded as identical phenomena,* and all explained in terms of a common cause, namely the absence of airy
and fiery little spirits which, like air and fire, always lighten
everything. We have here a self-contained, logical system built up
on a kind of analysis of feelings—at least on an identity of feelings.
Yet it is a system completely unlike our own. Just as we do, these
people observed, pondered, found similarities, and associated
them. They set up general principles, and yet they constructed a
system of knowledge completely different from our own. The
"heaviness" in this last example is a concept totally different from
that of our physical "weight." Many such examples could be given
all of which go to prove that comprehending objects and phenomena in a way similar to our own was completely alien to their way of
thinking. Our physical reality did not exist for them. On the other
hand, they were prepared to regard many another feature as real
which no longer has meaning for us. Hence we have these symbols,
parallels, profound comparisons, and astonishing statements.

To compare what for us is an alien and earlier thought style with
that of modern science, medical papers, especially anatomical
or physiological ones, are very suitable. These are easier for us to
comprehend than early physical or chemical papers, which for us
have become highly unintelligible.

I have before me on my desk a book about urine by Joseph Low,
M.D., dated 1815. The author was not one of the champions of
today's thought style. The book is steeped in the spirit of eighteenth century *Naturphilosophie.* We read, for instance: "The

---

*German *Schwere*, "heaviness," "weight"; *Schwermut*, "heavy-heartedness";
Schwierigkeit, "difficulty."—Eds.

manifestation of life is perfected only through its own creation. Life itself is only procreation and creation. The visible and perfected image of this continuous inspiration is the organic body, as the basis of that image.... For it is through the most intimate intercourse with life that organic matter receives life in its fullness. It thus perfects itself to become that first, primeval, and universal substance which is both procreating and birthgiving, which the ancients called prime matter and which is known by the moderns as nitrogen, azote, or more widely still as phosphorus" (p. 10). "Production of urine as a liquid in the human body corresponds to the formation of bone as a solid. Straddling the line dividing the inorganic from the organic forms, phosphorus becomes metallic as it returns to the liquid state in the urine together with all products accompanying it in bone formation. It thereby causes the substance even in the osseous system to change. The process of urine and bone formation therefore develops in only two directions and both meet at some stage of development throughout all the animal classes" (p. 41). "The quantity of phosphoric acid increases with age. And urea becomes uric acid, a spirit which is found only in human urine and indicates the perfect 'animalness' of man" (p. 56).

Far from being a pioneer, Löw is clearly a straggler. Phlogiston (p. 128) still haunts his book, and his concept of weight [*Schwere*] is quite out of keeping with the spirit of his time. ".... just as the stillness of the dead represents a sinking back into the metallic world, and when a person dies his body becomes heavier [*schwerer*][18] or metallic" (p. 43). Nevertheless, his thought style can be compared with the modern one, because many details in his book are directly comparable with details found in modern science. Löw considers himself a rational scientist and condemns the fanciful uromancy of the Middle Ages (p. 246): "Not until the sixteenth century, when the fanciful and extravagant uromancy of the Arabs aroused suspicion, did scientists return to the simple natural observation of urine." He thus regards his own theory as simple observation of nature, in just the same way as many contemporary scientists regard theirs.

Phosphorus is a rather fundamental idea for his chemical observations. But it is far from identical with the modern element of the same name, although there are some undeniably common features. "In all these natural properties of urine, phosphorus as the consummate product of the animal life process is the dominant, truly inspirited principle found in those many salts with an alkaline, nitrous base in the animal gluten, where it appears in a mucous, gelatinous state, in its primeval life-bearing predisposition either as decomposed basis of food or as element in the production of the first vegetable-animal beings, and in benzoic acid and hydrochloric acid as constituents thereof" (p. 12). "It is precisely the phosphorus in the urine which, during urine retention and as a ferment of death, rapidly induces a transition from a state of inflammation to gangrene. Its presence in urine may also stimulate the formation of phosphorus throughout the entire organism as, for instance, in urinary fever, which is a putrefactive fever of the worst kind caused by prolonged urine retention. That the meteoric-electric phenomena of the atmosphere...wield their enormous influence on the urinary system is also due to phosphorus" (p. 12). "Phosphoric acid is always augmented in carnivorous animals and predators.... The production of characteristic olfactory substances in the scent of such animals as well as the greater animality of the perspiration of mainly meat-consuming persons is very closely connected with more animalistic urine and especially with the augmented production of phosphoric acid in it" (p. 27). "Because of the increased quantity of urea, phosphoric acid, and the urinary salts, a copious crystalline precipitate occurs in the urine of the male as well as of the female, who always remains more true to the primeval life-conceiving state, and accordingly her phosphorus formation is even more gelatinous, soapy, oily, and greasy" (p. 44). "Phosphoric acid, called by Gärtner phosphorous acid, because it often causes urine and some types of sweat really to phosphoresce, is the only free acid in the urine" (p. 63). "The fibrin of the blood...represents the metallically formed phosphorus" (p. 100). "The increased production of phosphorus in inflamed urine is unmistakable in the color, temperature, consistency, quantity, and quality of the urine" (p. 115). "The two acids, uric acid and phosphoric acid, do not occur at all in urine from a nervous condition, because these express the inspiriting principle directly. But this must necessarily be absent here and cannot be generated,

because of the inhibited influence of the nerve spirit" (p. 157).
"The phosphorus in urine as uric acid, and as constitutive of the
urate, in addition to all the other salty and earthy qualities. . .
follows this urge" (p. 206). "Most of these urinary salt-concretions
have phosphorus as uric acid for a basic constituent" (p. 206).

Modern science has no term that properly describes this *phosphorus*. It is a principle, an axiom, and a symbol of the inspiratory,
animalistic as well as of the lethal forces "as the symbol of death."
It is related to meteoric electricity, the production of characteristic
olfactory substances, phosphorescence, inflammation, and putre-
faction. Like a chameleon it appears in such various guises as
metallic, gelatinous, soapy, or oily. It manifests itself as uric acid
and as a pigment, it forms concretions, and it gives urine its
color, temperature, and consistency. It is a principle, and yet it
becomes materialized in precipitates together with salts. It is ponderable, can increase or decrease and even disappear altogether. It
therefore does not share the properties of a modern principle,
because principles and symbols are now considered imponderable.
Yet it shares some features in common with the modern concept of
phosphorus. In particular there are some phenomena and connections which can be organized around either concept. Phosphorescence is predominant followed by high inflammability, the
ozone smell in the presence of phosphorous, and the same residual
smell following some "meteoric, electric phenomena," and then it
is present in large quantities in urine, in bone, and in the nervous
system. There is no doubt that some relation exists between the
modern scientific concept of phosphorus and that of Löw. Just
what it is, however, is difficult to describe in strict scientific terms.
It would be well to borrow the word "motif" from the field of art
and speak of an *identity of some motifs of both configurations*.
Both the source and the special relation to fire and smell would
thus be common motifs, which occur both in Löw and in the
modern scientific concept of phosphorus.[19]

These same attributes exhibited by phosophorus à la Löw—half
principle and half substance in a contemporary scientific sense—are
also exhibited in his other substances such as metal, water, and
urea. This gives his science a special stamp. Principles unite to form
moving [*pathetischen*] ideas, grandiose correlations and compari-

sons. In his kind of reality everything has the value of a symbol, which has an outer and less important form and an inner and profound meaning. The aim of his exploration is not to unearth and simply to explain this meaning but to divine it as a profound secret. We read, for instance, that "the kidneys, which developed from the mucosa of the genitals, have a special hidden relation and sympathy with the sexual system" (p. 43). "But it is procreation, the preparation of that all-procreative and birth-giving substance in the life process of the sexual organs, of phosphorus in its highest form of exaltation and inspiration, that connects the sexual and the urinary system in a way that is as profound as it is mysterious" (p. 44).

The deep mystery that the author finds here does not amount to a puzzle to be solved or a relation to be revealed by research. Conversely, the knowledge acquired about the relation consists only in *its being declared to be profound and mysterious.* It is the experience of mystery as mysterious. The awe he experiences when he looks at the veiled Isis is the intellectual gratification the author seeks, and what satisfies him.

When Löw finds what we would call a purely mechanical connection in some case, he is not satisfied and he looks for a deeper one. "During general paralysis of all the organs governed by the will, those of movement, the sarcoderms and the sphincters, all excretions occur spontaneously, simply because of the weak dilution of all humors." And urinary incontinence is only the expression of general "colliquative profluxia," which manifest themselves in the blood, making it "diluted, discolored, black, and foaming," as well as in the form of bloody sweat and diarrhea. He only has to look at an object to notice at once and to describe the profound, mysterious connections. "Just as the skin of the seriously ill produces a highly nitrous, cadaverously-smelling, contagious atmosphere, the urinary system produces turbid, black, foaming urine from which a black, sooty deposit resembling coffee-grounds soon precipitates giving off a putrid, stinking odor" (p. 111). Note the congruence in the dark coloring given to the details of this aspect. A "black" pessimistic prognosis corresponds to black urine; and with danger of death and contagion goes a cadaverous and putrid stench. This is not simply emotional fantasy. What we are faced with here is an explicit parallelism between the properties of the symptoms described and the meaning of the entire aspect. It is as if every part in concord betokened the meaning of the whole entity. The black color of the urine betokens the pessimistic prognosis, just as "the color, temperature, consistency, quantity, and quality of the urine" immediately and "unmistakably" betoken "the profuse production of phosphorus." Löw is ready to see such *signatures*[20] of profound meaning everywhere. He mentions a list of "indications of the urine" in the context of chronic diseases [pp. 140–41] and also of a "biliary signature of the urine" [p. 146]. These very signatures imprint the character of symbols upon the objects of his reality.

When reading his descriptions we are immediately struck by the descriptive terms he uses that are foreign to us. On page 120 we read about "ichorous fluid with mummy-like corrugation" into which the humors of a gangrenous organ degenerate. On pages 142 and 146 he writes about "jumentous urine" to describe a visual resemblance to the urine of grass-eating animals. His reports specify far too many qualities, which seems pleonastic to us. "The presence of synochal fever also becomes conspicuous by means of the urine through its discoloration and turbidity as well as through a lack of coction [*Mangel an Kochung*] and of homogeneous mixture. Its appearance is crimson or dark red, turbid, gluey, and it is rich in a diffuse, floccular, multicolored, dirty-white, often gray, mucous deposit, which consists of decomposed mucus, gluten, urea, and phosphorus. After sedimentation it always remains turbid and opaque, being already opaque when excreted; the urine retains this aspect right to the last stage common to all fevers" (p. 107). It is interesting to compare this profuse description with a modern account of the same condition. "Turbid, crimson or dark red urine with floccular deposit" is sufficient. Everything else is either useless to us (discoloration, of turbid complexion, gluey appearance) or has been replaced by the microscopic investigation of the deposit (the complex specification of the deposit). Statements such as "lack of coction and of homogeneous mixture" are wholly unfamiliar. The significance of these expressions, however, can be explained. They correspond to a pathological theory according to which all diseases progress in definite stages. The first is called the "stage of crudity" (*cruditas*). Corresponding to this stage is a "crude" urine which is "thick, turbid, discolored and shows no homogeneous mixture." It is characteristic that Löw should have listed "lack of coction" among the visible properties of the urine he describes. Although he construes such a lack as a directly visible property, we no longer regard it as such. It is a theoretically constructed gestalt*—*which Löw saw immediately, but which we do not.* Many other descriptions that are alien to us, such as the "jumentous urine" of grass-eating animals mentioned earlier, correspond to theoretically appropriate ready-made *gestalten, which we do not see* but which Löw, possessing the relevant stylized readiness to perceive them, *perceived directly,* analogously to the gestalten and qualities in present-day knowledge that are immediately perceivable without further ado, as we discussed in section 2 of this chapter.

All in all, Löw was ready to see features different from those we see and to convert what he saw into a different kind of cognition. To avoid any misunderstanding it must be added that Löw was certainly no great light among his contemporaries. He cannot even be regarded as their typical representative. My sole purpose was to cite an example of scientific thinking differing from that of today. In his particular mood, which to us seems fanciful and mystical, Löw is ready to see mysterious, deep connections. Furthermore, the objects of his world have a specific, symbolic coloring. This is his own particular *thought constraint, which becomes intensified until he directly perceives the appropriate gestalts.* At the same time, he considers himself to be a rational researcher, because after all he is only describing what he sees.

To obtain an even clearer picture of how scientific observation differs when two different thought styles are involved, it is perhaps appropriate to compare anatomical descriptions and illustrations in early and recent textbooks. I browsed through several seventeenth- and eighteenth-century anatomical textbooks, all of which provide almost equally suitable examples. Let me cite the description of the collarbone (clavicle) from Thomas Bartholin, *Anatomy, Old and New Observations, Especially the Teaching of*

---

*Where the alleged agency for properly homogenizing the mixture was missing.—Eds.

my Father, Caspar Bartholin, about Harvey's Blood Circulation and Lymph Vessels, fourth edition, Leyden, 1673.

Clavicles are called keys because they lock the chest, and like a key also lock the shoulder blade to the breastbone, or because they recall the keys to houses of ancient times, noted by Spigel at Padua in ancient houses. Celsus calls them jugula because they connect. Others call them tongues, the forked bone, or the upper support. They are located transversely below the lower neck, in the highest part of the chest, one on each side. They are shaped like an elongated Latin letter *S,* that is two semicircles sigmoidally joined, convex towards the outside joint and slightly hollow so that no vessels, which are large there, are compressed. In the male they are more curved so that the movement of the arms is less obstructed. In the female they are less curved to enhance her beauty, so that the depressions in this area are less conspicuous in the female than in the male, wherefore she is less proficient in throwing stones. The material is thick, but perforated and spongy. It is therefore often fractured, but easily knits together again. The surface is rough and uneven. They are connected to the upper process of the shoulderblades by cartilage, which joins them so that movement of the shoulderblades and arms is not restricted. But they are immobilized by ligaments surrounding the joint, with a broad and longish end, and joined to the breastbone at the other end as previously described. The collarbone is utilized for various movements of the arm, and because it is fixed by a bone like a stake, it can be more easily moved backward and forward. Hence animals except monkeys, squirrels, mice, and hedgehogs have no collarbone. [P. 745]

This description consists of: (1) a linguistic analysis of terms, taking up one-fifth of the chapter; (2) a brief description of the arrangement and a fairly detailed description of the connection with other bones; (3) a very graphic description, although poor in detail, of the shape; (4) a very brief description of the surface ("rough and uneven") and of the internal structure ("thick, but perforated and spongy"); (5) some comparatively comprehensive and very detailed teleological remarks, taking up about a quarter of the entire exposition; (6) and some brief remarks on comparative anatomy, such as, "Hence animals have no collarbones."

Let us compare this with a modern description, for instance the one under the heading "clavicula, clavicle" by Möller and Müller in a very concise compendium of anatomy.[21]

Concerning the *S*-shaped bone inserted between shoulder blade and breastbone, we speak of the middle section and the sternal and acromial ends. The middle section has an upper and a lower face with a shallow groove (musculus subclavius), an anterior (mm. pectoralis major and deltoideus) and a posterior edge. The sternal extremity is prism-shaped with anterior (m. pectoralis major), posterior (m. sterno-hyoideus), inferior and median faces (=facies articularis sternalis), superior (m. sternocleidomastoideus), inferior and posterior edges. The costal tuberosity is situated on the inferior face (ligamentum costoclaviculare). The acromion has a superior, inferior, and lateral face (=facies articularis acromialis), an anterior (m. deltoideus) and a posterior (m. trapezius) edge. The coracoid tuberosity (lig. coracoclaviculare) is on the inferior face. Development: main core is in the middle section with epiphysis at the sternal extremity.

Compared with the seventeenth-century description, the following changes will be noted. *No trace remains* of (1) the pseudo-linguistic analysis of nomenclature, (2) most of the vividness in the description of shape and arrangement, and (3) the teleological observations. *On the other hand,* (4) detailed information about muscle and ligament connections with the bones is provided, and (5) the description of the surfaces, the edges, the various parts of the bones is far *more detailed.* The shift in intellectual interest is very clear. What Bartholin described in just a few words has become ten times as detailed, but what he described in great detail has almost disappeared. In the place of the nomenclature analysis and teleological observations, constituting almost half his text, detailed connections of bodily organization are now described. Personal names as well as any popular aspect of form and purpose have been relegated to the background by a detailed description of connections in terms of a mechanico-technical theory.

The characteristics listed here can be found in all early anatomical descriptions, often in a style even more pronounced and gross. There are nomenclature analyses occupying half a page, with citations, discussions, deductions, and opinions. In an epitome of Vesalius' *Anatomy,* edited by Fontanus,[22] the chapter on the thighbone (femur) devotes only 31 words to its anatomical structure, in the modern sense of the term, compared with 135 to a description of the name "femur" and its meaning in Pliny, Plautus, Virgil, Horace, etc. In Bartholin we read, for instance,

"The belly-as-stomach is called as it were a little belly-as-abdomen" (p. 66), or "testes or testicles attest to a man's virility" (p. 208), or "The ticker-as-heart is so called in virtue of its ticking motion" (p. 353). A name here has a completely different significance from what it has today. It is not an arbitrary, conventional designation or one that arose by historical accident. The meaning is inherent in the name, and its investigation constitutes an integral part of acquiring knowledge about what it names. The name ranks as a property of its object of reference.

Early anatomical descriptions and illustrations are characterized by their graphic quality. We noted this in the description of the collarbone. Bartholin wrote about the kidney as follows: "Its shape is that of a kidney bean, or of a liverleaf when looked at in profile. Outside, the shape is gibbous and round in the back as well as towards the abdomen. The inside upper and lower parts are gibbous in shape, but the middle section concave and snubnosed" (p. 177). Books on anatomy from the seventeenth and eighteenth centuries contain absolutely superb graphic pictures of nerve men and vein men which can never be found in modern textbooks. But this clarity has a distinct coloring. Figures of skeletons, for instance, do not just illustrate bones, or even a systematic arrangement of the bones, but express an emotive symbolism. They symbolize Death by carrying spades, scythes, or other insignia of death.[23] Figures of muscular men are represented as martyrs. Other figures also assume pathetic postures. Faces do not exhibit the empty expression of corpses or the diagrammatic features typical of modern anatomical illustrations, but are expressive and distinguished. In the representation of an unborn child, both the proportions of the fetus and the position of the limbs are arranged in a conventional, amoretto-like way. The head is much too small and the limbs assume a comely position not corresponding to the compact position of the embryo.[24] If we look at the earliest anatomical illustrations, such as the accompanying illustration from the twelfth century, the first feature that strikes us is their schematic and primitive symbolic character. We see figures set in conventional uniform postures, the organs are indicated symbolically, such as the circular duct in the thorax, meant to represent the circulation path of the pneuma in the chest, and below on the right the schematic five-lobed liver. What we have here are ideograms [*Sinnbilder*] corresponding to then-current ideas, not the form which is true to nature as we construe it. Intestinal loops, for instance, are not portrayed as a certain number of sections positioned in a certain way but as spiral lines symbolizing the loops (see illustration). Nor do we see definite convolutions of the brain but the "curliness of the brain surface in general"; not a certain number of ribs but the "ribbing of the chest wall in general." The cross section of an eye does not reveal a definite number of wall layers but its multilayered structure schematically represented, which makes the illustration resemble a cross section through an onion.

We are thus confronted with ideograms [*Ideogramme*], or graphic representations of certain ideas and certain meanings. It involves a kind of comprehending where the meaning is represented as a property of the object illustrated.

The very detailed teleology, endeavoring to find a meaning in every detail, is perhaps connected with such ideovision [*Sinn-Sehen*]. The book of Fontanus (p. 7) contains such a description. "The lower ribs are shorter, so that the full stomach is not compressed, and they are more pliable for the same reason." The bone sutures of the crown of the skull have the purpose of releasing "vapors" from the skull (p. 3). That the fingers each have three phalanges, that the cartilaginous rings of the trachea are not completely closed, etc. are further details each allotted a simple, as it were, primitive purpose.

The interpretation of the anatomical illustrations as ideograms* forces itself all the more upon us the more alien the author's thought style and the further removed from us the era concerned. All we see in medieval, in Persian, and in Arabic illustrations is schematic sign language but almost no realism.[25] The difference between one of these alien thought styles and the modern one does not rest simply on our greater knowledge. They have actually more to say about that which in their particular reality has a greater value than it does in ours. Bartholin has also written a chapter on the sesamoid bones

---

*Here Fleck uses *Sinnbilder,* followed by *Ideogramme* in parentheses, thus identifying in this context the two terms with one another.—Eds.

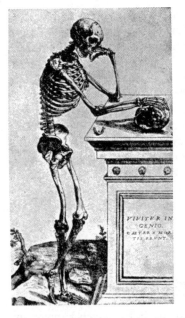

Vesalius' drawings of skeletons. After Roth 1892.

The rib cage. After Heitzmann 1888.

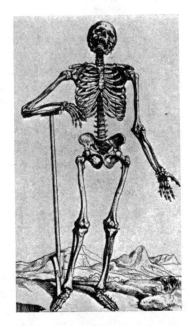

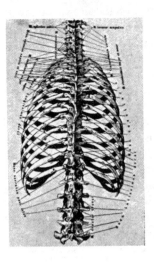

Figure 3

(p. 756). This is even longer than the chapter on the cervical or collar muscles and consists of about thirty times as many words as the few that are included in modern anatomies about these bones [or superfluous cartilaginous nodules].[26]

These [sesamoid] bones are important in Bartholin's osteology but without importance in ours, standing as it were beyond the pale of the osseous system altogether. Bartholin still subscribed to ancient, fanciful legend according to which these little bones are seeds from which bodies will one day again grow "like a plant from its seed." He did not believe very firmly in it, but he nevertheless felt obliged to cite the other authors, discuss the purpose of the bones, deal with their form and position, show surprise at the variability of their number, etc. Accordingly he had more to say about them than we do, and even more than about the cervical muscles, which today constitute an important part of myology.

He wrote almost five pages about the hymen, which today is described in one or two sentences. A great deal of space in these old descriptions is devoted to counting the number of anatomical parts. Fontanus notes: "There are twenty bones in the skull, of which eight are in the head and twelve in the upper maxillary" (p. 36). He tells us that there are twenty-eight bones in the toes, and the total number of human bones is 364; that there are seven pairs of muscles which move the eye and four pairs the cheeks and lips; that the portal vein forms five branches, etc. Today such counting is impossible, since we often regard it as arbitrary whether three bones or four, for instance, can be separately identified in a given articulation. But thought styles exist in which the number, just like the name of the object described, is important not as a means of description but in its own right. Only a vestige of such number mysticism remains with Fontanus. But in many thought styles such as the Indian thought style and that of the Chinese, such a system was elaborated until it formed a rich number cabala, in which numbers were accorded special signification and meaningful connections were established among them. If a thought style is so far removed from ours as this, no common understanding is any longer possible. Words cannot then be translated and concepts have nothing in common with ours. Even shared motifs such as the

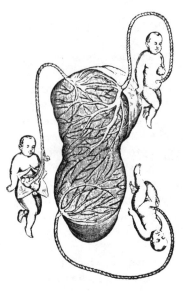

Triplets. From Bartholin 1673.

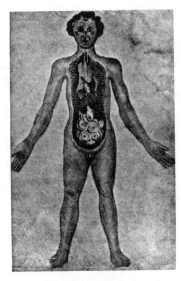

Bloodletting figure, fifteenth century. After Sudhoff.

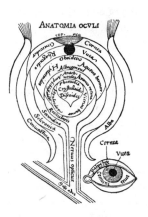

Cross section of the eye, 1539. After Sudhoff.

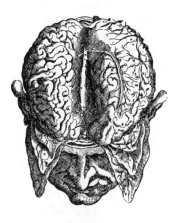

Surface of the brain. From Vesalius 1543.

Figure 4

affinity exhibited between Löw's concept of phosphorus and the modern one are missing.

To the unsophisticated research worker limited by his own thought style, any alien thought style appears like a free flight of fancy, because he can see only that which is active and almost arbitrary about it. His own thought style, in contrast, appears imperative to him,* because although he is conscious of his own passivity, he takes his own activity for granted. It becomes natural and, like breathing, almost unconscious, as a result of education and training as well as through his participation in the communication of thoughts within his collective. Modern anatomists would regard as a useless emotional frill any representation of the skeleton as a symbol of death, such as was typical of Vesalius himself as well as of his predecessors and contemporaries. But we can learn to see their particular intellectual mood even in our present-day anatomical illustrations. Consider, for example, figures 120 and 121 of Heitzmann's anatomical atlas, which represent the rib cage.[27] A mechanico-technical cage motif is in keeping with this representation, just as much as a salient death theme with the skeleton figures of Vesalius. It cannot be claimed that the resemblance to a cage arises "automatically." It appears only after (1) a purposeful stripping of the ribs, (2) a purposeful assembly of the plexus, as well as (3) a purposeful arrangement of the whole to bring about this resemblance in perspective, in a manner analogous to the purposeful ideography [*Aufstellungen der Sinnbilder*] of early anatomy. Furthermore, (4) the lines added to indicate muscular insertions underscore the symbolism of a mechanical apparatus every bit as much as the scythe underscores the symbolism of Death for Vesalius. *These modern figures are ideograms just like those of Vesalius.* There is no visual perception except by ideovision and there is no other kind of illustration than ideograms.

A technico-mechanical motif is in keeping with all osteological figures of modern anatomy. Accordingly the skeletal system is regarded as a supporting frame. Everyone is so familiar with this idea both from school and from our thought style that we are

---

*"Der eigene Denkstil erscheint ihm dagegen als das Zwingende," in the original, implies that the individual *can be aware* of the coercive function of his own thought style. Cf. p. 41—Eds.

forced to exclaim that of course "it really *is* the supporting frame."
It certainly is, provided we are thinking according to the thought
style of modern science. But it is not difficult to imagine a system of
knowledge in which the skeleton is not construed as a frame
supporting the body. If one adheres to the concept of heaviness
[*Schwerebegriffe*] found in Schreger and even in Löw, for instance,
it is by no means impossible to look to the airy and fiery spirits as
supporting the body, because these keep the body upright by their
urge to rise. Here the bones would really be the opposing element,
which is lifeless "metallic" and non-"inspirited." "As all persons,
when they die, become heavier or metallic..." As the non-in-
spirited principle of the body, and mere ballast, the skeleton would
attract much less attention and be depicted as a pile of bones
rather than the frame shown in modern anatomical illustrations. In
about the same way, fatty tissue appears in modern anatomical
illustrations not as a continuous system but as a kind of photo-
graphic negative. It "appears" as that which has been cut away
[much like the "lack of coction" which Löw "observed"].

We have defined thought style as the readiness for directed
perception and appropriate assimilation of what has been perceived. We have already mentioned the particular mood which
produces this readiness for any particular thought style. An exhaustive investigation of thought styles cannot be assigned to this
book, for it would take up the working capacity of a lifetime. There
is but one element of the thought style of modern science that
ought to be discussed, namely the specific intellectual mood of
modern scientific thinking, especially in the natural sciences. This
mood stands in direct relation to the specific structure of the
thought collective of science as has already been described.

It is expressed as a common *reverence* for an ideal—the ideal of
objective truth, clarity, and accuracy. It consists in the *belief* that
what is being revered can be achieved only in the distant, perhaps
infinitely distant future; in the *glorification* of dedicating oneself to
its service; in a definite *hero worship* and a distinct *tradition*. This
would be the keynote of the common mood in which the thought
collective of natural science lives its life. No one already initiated
would claim that scientific thinking is devoid of feeling. Nor can
there be any doubt, according to our argument, that the particu-

Figure 5

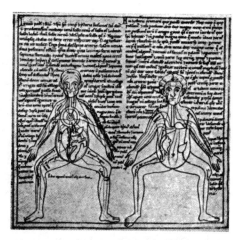

Anatomical illustrations, 1158.
After Sudhoff.

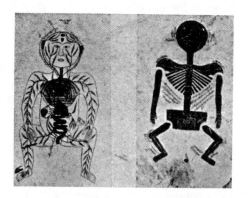

Skeleton figure, 1323. After
Meyer-Steineg and Sudhoff
1928.

Persian anatomical
illustrations. After Meyer-
Steineg and Sudhoff 1928.

Fetus in utero, about 1460.
After Weindler 1908.

lar attitude influences not only the work method but also the results. It manifests itself concretely as a readiness for directed perception.

But how is this mood put into effect? First, every scientist has the obligation to remain in the background. This obligation is also expressed in the democratically equal regard for anybody that acquires knowledge. All research workers, as a matter of principle, are regarded as possessing equal rights. And all, in the service of the common ideal, must equally withdraw their own individuality into the shadows, as it were. Personal supposition in science is regarded as provisional; this is a specific structural aspect of the thought collective of science. We previously discussed in detail the centrifugal tendency of the products of scientific thought [*der naturwissenschaftlichen Denkgebilde*] and the centripetal feedback of this tendency in the form of a migration of ideas throughout the collective between the esoteric and the exoteric circles. We emphasized the distinctive "modesty of the plural" as well as a characteristic personal modesty and caution.

The mood of the thought collective of natural science is further realized in a particular inclination to objectivize the thought structures [*Denkgebilde*] that it has created. This is the counterpart to the obligation of the scientist to withdraw as a person. This tendency to reify and objectivize the conceptual creations of scientific thought [*Denkgebilde*] arises, as has already been described, during the migration of ideas throughout the collective and is inseparably bound up with it. Graduated in several steps, it begins with statements by different scientists as well as the historical development of a problem, so that it becomes depersonalized. Special expressions or "technical terms" are introduced. To these are added special symbols and possibly a whole sign language such as is used in chemistry, mathematics, or symbolic logic. Such a lifeless [*lebensfremde*] language guarantees fixed meanings for concepts, rendering them static and absolute. A further factor is the particular reverence for number and form as well as the striving for vividness and a closed system. A maximum of information is demanded, the greatest possible number of mutual relations between individual elements, in the belief that the ideal of objective truth is all the more closely approached as more and more relations are found.[28]

Thus, a structure [*Gebilde*] is created step by step. Starting as a unique event or *discovery,* as seen from the history of thought, this is developed by the extraordinary forces of the thought collective into what seems to it to be a necessarily recurrent and thus objective and real *finding.*

The disciplined, shared mood of scientific thought, consisting of the elements enumerated, connected with the practical means and effects, yields the specialized thought style of science. Good work done according to style, instantly awakens a corresponding mood of solidarity in the reader. It is this mood which, after a few sentences, compels him to regard the book highly and makes the book effective. Only later does one examine the details to see whether they can be incorporated into a system, that is, whether the realization of the thought style has been consistently achieved and in particular whether procedure has conformed to tradition (= to preparatory training). These determinations legitimize the work so that it can be added to the stock of scientific knowledge and convert what has been presented into scientific fact.

Commentary and Annotation

Biographical Sketch

Ludwik Fleck was an unusual man, a humanist with an encyclopedic knowledge. He had a very keen sense of humor and was erudite in many fields. In his person were united the excellence of a distinguished microbiologist and the depth and insight of a great philosopher.

Fleck was born in Lvov, Poland, 11 July 1896, to Maurycy Fleck and Sabina (neé Herschdörfer). He attended grammar school and high school in Lvov, where the German language still dominated from the period when Lvov, as Lemberg, had belonged to the Austrian Empire. In 1922, at the age of twenty-six, Fleck received his medical degree from Lvov University. From 1920 to 1923 he was assistant to Dr. Rudolf Weigl, well known for his research on typhus, both at the Typhus Investigation Laboratory of Lvov University and at the State Hospital for Infectious Diseases. He later specialized in bacteriology in Vienna. From 1925 to 1927 Fleck was head of the bacteriological and chemical laboratories of the State Hospital in Lvov. He spent the year 1927 in Vienna (during the heyday of the Vienna Circle), working in the State Serotherapeutic

This sketch is based on correspondence and documents from the Fleck estate provided through the kindness of Mrs. Ernestina Fleck and Professor Marcus A. Klingberg, Israel Institute for Biological Research, Ness-Ziona, Israel.

Institute under the direction of Dr. Kraus. From 1928 he was head of the bacteriological laboratory of the Social Sick Fund in Lvov until his dismissal in 1935—an anti-Jewish measure.

By 1935 Fleck had already published extensively on various aspects of general serology, hematology, experimental medicine, immunology, and bacteriology. Among the specific topics of his research were the serology of typhus fever, syphilis, and a variety of pathogenic microorganisms. Before World War II, Fleck carried out extensive research on epidemic typhus fever. He developed an original method to obtain a vaccine and a diagnostic skin test, which he designated the "exanthin reaction." This was internationally confirmed and is mentioned in textbooks. Besides his specialized scientific research on cytoserology and infectious diseases (including the concept of infection), Fleck published seven papers on methodology of science, some articles on the methodology of scientific observation, on principles of medical knowledge, on the history of discoveries, etc. He also developed an original theory about thought style and thought collectives which he formulated in the monograph translated above. Completed in 1934, the work was published in 1935 in Switzerland (because political conditions in 1935 did not permit a Jew to publish in Germany) and was widely discussed in Poland, Germany, France, Italy, and Switzerland.

Between 1935 and 1939 Fleck worked in private practice at Lvov in his private microbiological laboratory, where he could also pursue his research. It was here that he continued his investigations into the variability of streptococci and the etiology of pemphigus vulgaris. He examined the role of normal sera on the course of serologic reactions, found a new method to strengthen the sensitivity of the Wassermann reaction, and discovered an original method for distinguishing true serological reactions from pseudoreactions. In December 1939, after the Russian takeover of Lvov (it is today just within the Ukrainian Republic's borders), Fleck was made director of the City Microbiological Laboratory. At the same time he was appointed to the teaching staff in the microbiology department of the State Medical School in Lvov. Until 1941 he also served as head of the microbiology department of the State Bacteriological Institute in Lvov as well as consultant in immunology and serology at the State Institute of Mother and Child Welfare in that city.

During the German occupation of Lvov from 1941, Fleck was director of the bacteriological laboratory of the Jewish Hospital there until the following December. Concerned with the severe typhus epidemic which arose under the ghetto conditions, he succeeded with very primitive means to develop a rapid diagnostic test for typhus, which permitted early detection and isolation. The excretion of a specific antigenic substance from the urine of patients with typhus fever was used for these diagnostic and vaccination purposes. Fleck used the urine of typhus patients as a source of the rickettsial antigen which proved to be a very effective vaccine. It was not until 1947, however, that he was able to publish the results of this new method of typhus immunization. In 1942 the Germans arrested Fleck and his family, and forced him to produce his vaccine for the German armed forces (he was asked by the German commanding officer whether such a vaccine prepared from urine would be suitable and useful also for Aryans. Fleck answered: "Of course, provided that the vaccine is prepared from the urine of Aryans and not of Jews").[1]

After Fleck was forced to divulge to several German doctors the procedure for obtaining the new vaccine, he and his staff were deported to the Nazi concentration camp in Auschwitz. In 1943, after recovering from a serious illness, Fleck was attached to the camp's hospital and, under duress, continued to produce his vaccine for the German armed forces. His two sisters, Dr. Henryka Fleck-Silber and Antonia Fleck-Kessler, along with their families, were killed by the Nazis in Poland. In 1944, Fleck was transferred to the concentration camp in Buchenwald and again ordered to prepare typhus vaccine. When this camp was liberated by the United States army on 11 April 1945, Fleck returned to his native Poland, where, from October of that year, he served as assistant professor of microbiology and head of the Institute of Microbiology at the newly founded Marie Curie Sklodowska University of Lublin. He organized a microbiological institute and, in August 1947, became associate professor of microbiology.

In 1947 Fleck discovered a new phenomenon occurring during inflammation: a clumping of white blood cells, which he designated "leukergy" and to which he devoted a great deal of his research over the subsequent decade. About forty articles were devoted to this effect by Fleck and his students. The discovery had an immediate impact within the scientific community.[2] In 1949 Fleck received the scientific prize of the city of Lublin for his research on leukergy.

During the 1948 Nuremberg trials, Fleck was invited to participate by the American authorities, because of his expert knowledge. He "cooperated and rendered substantial assistance to the prosecution in the case of the United States versus Krauch et al.," according to the Office of the Chief Council for War Crimes. From June 1950, Fleck was full professor of microbiology at Lublin until his appointment in 1952 as director of the Department of Microbiology and Immunology at the Mother and Child State Institute in Warsaw. In the following year he was awarded the state scientific prize of Poland for his research on epidemic typhus fever. He was elected a member of the Polish Academy of Sciences in 1954. In 1955 he received the highest scientific degree in Poland—a doctorate of medical science. That same year he was invited by Professor Trefonel of the Pasteur Institute in Paris, as well as by Professor Fontaine of the medical faculty in Strassburg, to lecture on leukergy. In 1956 he was invited to a conference on autoantibodies at the University of Texas.

In 1957 Fleck finally succeeded in realizing his wish to emigrate to Israel. His only son Ryszard (Arieh, in Hebrew) Fleck had lived there since 1947, working in the laboratory of the Labor Sick Fund in Petach Tikwa. Ever since the end of the war, Fleck had been trying to go to Israel, but it was 1957 before he was allowed to leave Poland in such a way that he could take his wife, Ernestina, with him. He joined the Israel Institute for Biological Research at Ness-Ziona as head of the Section of Experimental Pathology.

Fleck continued his research on diphtheria during the 1950s, developing a vaccine, "Anabac," which he presented to the public in 1957. After 1957 he developed some new serological tests for brucellosis and measles.

In addition to the monograph translated above and textbooks on hematology, serology, and bacteriology, Fleck published over 130 scientific articles, in Polish and German as well as in Hebrew, English, French, and Russian. He was a member of many international scientific societies, including the New York Academy of Sciences, the International Haematological Society, and the International Society of Microbiologists.

Fleck suffered from Hodgkin's disease and was seriously ill by the beginning of 1961. Nevertheless he was invited by Professor Nathan Rotenstreich of the Department of Philosophy, Hebrew University of Jerusalem, to present a course of lectures there. The offer pleased him, and he started to prepare the lectures, but failing health prevented him from delivering them. He died on 5 June 1961 and was buried in Ness-Ziona.

Descriptive Analysis

Fleck turns to the history of medicine—taking as his case study the concept of syphilis and the discovery of the Wassermann reaction—to document his general theory of the sociology of knowledge. While the notion of collective knowledge is not unique with Fleck, he seems to have been the first systematically to apply sociological principles to the origin of scientific knowledge. Fleck explicitly opposes those sociologists who consider science an inappropriate subject for their investigations. Taking the complex of history, philosophy, and sociology of science, he discusses such currently important issues as theory conflict and change, and the role of anomaly and error in scientific discovery. Fleck considers every concept and theory, whether scientific or not, to be culturally conditioned, including his own theory of thought style and thought collectives.

When syphilis first appeared in Europe in the fifteenth century, it was construed as a scourge brought about by sin. As methods of treatment emerged, the concept became associated with the method of cure. Gradually a more scientific concept emerged in conjunction with the causative agent of disease.

Syphilis was early associated with the blood. The idea of "bad blood" took on mystical-ethical overtones. Only with the discovery of the *agent* of the disease in the blood did the association become scientifically sound.

The association of syphilis with the blood emerged only gradually from a matrix of vague generalities and taboos. The development of a concept such as this is conditioned by cultural-historical factors which represent to some extent a thought style (*Denkstil*) characteristic of the given era. The discovery of the Wassermann reaction further defined and specified this concept. Although the preliminary experimental results were not reproducible, the Wassermann reaction became accepted because it soon proved to be extremely useful. The very difficulty of reproducibility, as Fleck points out, demonstrates the social dependence of all knowledge.

Fleck sheds light on the way ideas, concepts, and theories are shared by individual members of the scientific community. He regards the role of the individual in scientific discovery as subordinate, since each individual shares in, contributes to, and draws upon the collective for his experience and ideas. Comte, Durkheim, Lévy-Bruhl, and other sociologists were wrong, Fleck believes, to exempt scientific knowledge from sociology and uncritically accept accumulated progress in scientific knowledge, as if our way of thought represented an improvement upon the thought style of previous generations. He emphatically rejects the notion that currently recognized "facts" are more true, opposes the Vienna Circle by rejecting any absolute and objective criteria of knowledge, and challenges Carnap to discover for himself the social conditioning essential for scientific knowledge (see chap. 4, note 3, above). Carnap, Schlick, and other members of the Vienna Circle sought to free science from cultural and other influences so as to establish it on an absolute and objective foundation. Sociologists such as Jerusalem at Vienna explored as an alternative the culture conditioning and social dependence of scientific knowledge. In Germany this relativistic approach unfortunately developed into a nationalistic and racist movement, yielding the dual perversions of German physics and Jewish physics. Fleck clearly sympathized with the Vienna sociologists in opposition to the dogmatists, but was equally wary of understanding such cultural factors principally along nationalistic or racist lines.

The social and cultural conditioning of scientific knowledge for

Fleck seems to have been based on a quite different ideology; if socialistic, certainly not of the totalitarian variety. For Fleck, facts are a function of thought styles, and these vary in a nonprogressive way with time and culture. Truth is neither relative and subjective nor absolute and objective but essentially determined and measured by a given thought style. If the stated aim of the Vienna Circle was to attain a general structure of unified science by applying the particular method of logical analysis combined with symbolic logic to the empirical material, reduced to what is immediately given in experience, Fleck counters that there is no unconditioned experience, and that such a rigorous method is inappropriate, there being neither a fixed cognitive substratum nor a closed and final constitutive system of scientific knowledge.

Science for Fleck is not the exact theoretical structure it is often taken to be but must be approached relativistically with due consideration for sociological and axiological factors. There is no such thing as complete truth or error for Fleck. Indeed, there are many "correct" theories for the same problem. Truth in science is a function of the particular style of thinking that has been accepted by the thought collective. To be correct is rather to be accepted collectively. Thus truth can vary with time and culture, for it is determined by a given thought style. Fleck rejects as irrelevant any simple correspondence theory of truth relating words to objects or cognition to facts. There is no objective and absolute truth. Truth is rather a stylized solution which is unique and singular only with respect to a particular thought style. It is not so much subjective as intersubjective or collective. And although relative it is not arbitrary, since it is a function of thought style (see chap. 4, sec. 3).

Existence and reality are also relativistically construed as constantly in flux (see chap. 2, end). Reality for Fleck is simply a systematic harmony of illusion which is acceptable because coherent. Existence or reality is the *passive* aspect of knowledge. It is that which manifests itself in particular consequential results from specific assumptions and preconditions *actively* assumed by the collective. The active aspect involving the choice of a particular set of preconditions can be explained in terms of historical, sociological, and psychological factors. But the passive fixed reality, involving the notion that any attempt will meet either success or failure, cannot be accounted for in these terms alone. We may *actively* select 16 as the value for the atomic weight of oxygen, for example, but if so then the atomic weight of hydrogen "will inevitably be 1.008. This means that the ratio of the two weights is a passive element of knowledge" (chap. 4). But this passive aspect of knowledge depends partly upon the given set of assumptions actively made according to a particular thought style and hence reality is also a function of time and culture.

Fleck also stresses the sociological factor as being essential to knowledge in general and to scientific knowledge in particular. Science is conditioned by such factors and hence can be explained more adequately in terms of the sociology, history, and psychology of ideas. Every era has its own dominant opinions and every age its own thought style. Any kind of cognition is for Fleck a social process. What is unique and remarkable about Fleck's theory is the emphasis he places upon the relevance of this sociological dimension to scientific knowledge.

How Facts Arise and Develop

As it stands here, the title of Fleck's monograph presents a direct challenge to the alleged primacy of facts. Facts are not objectively given but collectively created. For Fleck there is no such thing as a fact which is impossible in principle. Any fact is possible as long as—indeed only if—it fits the accepted thought style. Facts, like ideas, arise collectively, spontaneously, and impersonally. Although Fleck does allow for individual exploits without overt dependence upon a collective, he stress that these can be successful only if the time is ripe for acceptance. There are no bare facts. Facts arise and are known only by virtue of the given thought style characteristic of a given thought collective. A fact begins with a tentative signal of resistance by the collective. This preliminary signal of resistance is but the predisposition for an emergent fact. Through collective interaction this tenuous indication gradually becomes stylized, undergoes consolidation, and emerges as an accepted fact. Such a fact does not stand alone but becomes a new feature of an interlocked system of ideas all of which are congruent one with another on the basis of a given thought style.

The development of knowledge consists not in a mere accumulation or increase of passive connections. Assumptions can also change, and thus there can be an increase in the active aspect of knowledge as well. Knowledge not only increases but, perhaps more important, undergoes change. according to thought styles. In Fleck's view, there is no end to possible systems of opinion, just as there is "no limit to the development of biological forms" (chap. 3). Thus our current opinions are also subject to revision. Fleck construes knowledge as involving not only a dialogue between the knowing subject and the known object but a threefold relationship, which includes the collective. Cognition, as a function of three components, is a relation embracing the individual subject, the particular object, and the thought collective within which the subject acts; and holding only under the conditioning influence of the particular thought style originating within the given cognitive community. "Between the subject and the object there exists a third thing, the community. It is creative like the subject, refractory like the object, and dangerous like an elemental power."[1]

Thought Styles and Thought Collectives

A thought collective is the communal carrier of a given thought style and of the stylized works of the mind produced thereby. Although such a collective consists of individual persons (chap. 2, sec. 4), it is not this substantive aspect that is crucial. For Fleck, a thought collective is to be construed principally in *functional* terms (chap. 4, sec. 3). To the extent that Fleck tends to personify the thought collective, he justifies this quasi existential status or hypostatized fiction as being pragmatically useful (sec. 4, n. 7). Since a thought collective serves mainly a functional role for Fleck, he construes individuals as constituting such a thought collective only by virtue of their *active* and actual exchanging of thoughts.

Fleck defines the thought collective as "a community of persons mutually exchanging ideas or maintaining intellectual interaction" (chap. 2, sec. 4). The meeting of minds of two persons is sufficient to constitute a thought collective if it satisfies the functional criterion (chap. 2, sec. 4; chap. 4, sec. 3). Besides such transient albeit repetitive thought collectives there are more stable thought collectives such as a nation, a race, a political party, or a social class. Thought collectives can be found in commerce, fashion, art, science, politics, the military, and religion. Fleck designates the thought collectives that are more permanent and stable as "thought communities" (chap. 4, sec. 3; chap. 2, sec. 4). He uses the term "thought collective" in a variety of ways to refer to a political party, to a small group of members discussing an issue, to a pair of such individuals, as well as to the concept itself.

What links the individuals of a thought collective together is the thought style they share. Considered in its collective function, the thought style is "the special carrier for the historical development of any field of thought, as well as for the given stock of knowledge and level of culture" (chap. 2, sec. 4). But Fleck defines thought style, insofar as it influences the individual, as the readiness for directed perception, with corresponding mental and objective assimilation of what has been so perceived (chap. 4, secs. 3, 5).

A thought style functions by constraining, inhibiting, and determining the way of thinking. Under the influence of a thought style one cannot think in any other way. It also excludes alternative modes of perception. Accordingly, no proper communication can arise between different thought styles. A thought style functions at such a fundamental level that the individual seems generally unaware of it. It exerts a compulsive force upon his thinking, so that he normally remains unconscious both of the thought style as such and of its constraining character. Yet such a style can be revealed in practice by an examination of how it is applied. The existence of stable thought collectives suggests the presence of a rather permanent thought style.

Several difficulties surround the notion of thought collective, thought style, and the relation between these concepts. A thought collective is construed functionally and as involving an actual exchange of ideas. Yet he describes as thought collectives both a political party, which hardly engages actively in the exchange of views, and a small group of party members, which does. Fleck does not seem clearly to distinguish between the thought style as a latent dispositional state giving an enduring character to thought collectives, and thought style as an active expression of a thought collective. He does point out that long-lived thought collectives tend to have permanent thought styles, but this is somewhat of a truism. And presumably such a thought collective persists during periods when there is no opportunity for active conversation and exchange of ideas. Related to this is the problem of which is prior, the thought collective or the thought style. Thought collectives seem to persist throughout periods of thought style revision. This point becomes acute in the case of the bold individual with a new but alien thought style. Fleck observes that such a thought style will require a long time to be collectivized, but it is not always clear whether this means that the new thought style must infiltrate the old thought collective or that a new collective forms around the new thought style.

Fleck lays such stress upon the collective that even the individual can be understood only in collective terms, although "the collective is composed of individuals" (chap. 2, sec. 4). Somewhat paradoxically he finds that the collective is more stable and consistent in its personality and existence than the individual. Yet the thought collective is more than merely the sum of the individuals (chap. 2, sec. 4; chap. 4, sec. 4). The individual seems to be necessary only to the extent that he provides sensory physiology and psychology. Creativity is a function of the collective, not really of the individual. If individual contributions are recorded in the history of science, these are considered exceptions which somehow lacked the relevant contributions from a collective. An individual must conform to the collective and to the extent that he may not, he is considered deviant. Yet it is remarkable that Fleck feels the need to include in an essential manner such individuals who possess a strong "personal thought style" (chap. 4, sec. 4). Such an individual forms a unique mono-collective as he conducts a dialogue with himself. He participates in more than one collective at the same time. As a "marginal man" he is crucial for the exchange of thoughts between different thought communities (chap. 2, sec. 4; chap. 4, secs. 3, 4).

The Thought Collective and Thought Style of Science

Fleck divides the thought collective of science into two concentric circles: an esoteric circle of experts, surrounded by the large exoteric circle of laymen. The esoteric circle has a hard core of special experts within the circle of general experts. There is also a stratification of the exoteric circle including a more practically oriented circle lying closer to the esoteric. An individual can pass from the exoteric circle to the esoteric by undergoing a process of initiation in the form of scientific education. There are accordingly four types of science each with its own characteristic literature: (1) journal science for the special experts, (2) vademecum or handbook science for the general experts, (3) popular science for the exoteric circle, and (4) textbook science for initiation into the esoteric circle. The most characteristic operational feature is a democratic exchange of ideas and experience, going outward from the esoteric circle, permeating the exoteric circle, and then feeding back upon the esoteric circle. The work of the mind thus conveyed undergoes a process of social consolidation and becomes thereby a scientific fact.

The thought style originates within the esoteric circle and is communicated outward throughout the entire exoteric circle. Understanding the work of the mind produced according to the thought style is greatest within the esoteric circle, because of initiation, and this decreases centrifugally. At the same time the degree to which the thought style exerts a coercive influence upon thinking increases centrifugally, becoming so dominant at the periphery of the exoteric circle that it is taken for granted. The esoteric circle is influenced by its own thought style once this takes on a collective character. The thought style and the ideas it produces become generalized as they are communicated throughout the exoteric circle. In this generalized collective form, these feed back centripetally upon the esoteric circle, producing a tenacious, self-contained, certain, impersonal, and collective system of thought. In an advanced state of scientific development there is accordingly less room for differences of opinion within a given thought style.

The thought style and thought collective of modern science can be construed in at least two ways. More restrictively, they can refer to scientists and related individuals sharing the thought style of modern science. But Fleck sets no clear boundaries to the exoteric circle, and on occasion seems to include the general public. Less restrictively, then, thought style can be taken as a cultural viewpoint. In this sense, the thought style of modern science becomes a cultural trait—the scientific thought style of Western thought.

In addition to the intracollective communication of ideas from the esoteric circle throughout the exoteric circle and the feedback of the ideas, reinforced and collectivized, Fleck notes an intercollective exchange. This complex process between collectives having different thought styles leads to the deterioration of fixed systems of opinion. The *change* of thought style opens up "new possibilities for discovery and creates new facts. This is the most important epistemological significance of the intercollective communication of thoughts" (chap. 4, sec. 3).

Where there is no feedback from the exoteric circle, as in the thought collective of religion, the esoteric elite exerts a dominant and dictatorial influence upon the masses. In contrast, the esoteric circle of the thought collective of science, with its symmetrical exchange, is democratically dependent upon public opinion from the exoteric circle. The leaders within the esoteric circle may provide the directions for scientific progress, but it is the collective that decides which innovative path is to be followed. "The thought style of natural science . . . denies previous knowledge any preferential or privileged status above that of new knowledge" (chap. 4, sec. 5, note 28).

For Fleck, all empirical discovery can be understood as either a supplement, a development, or a transformation of thought style. Once a thought style takes hold, any exception to its system initially appears unthinkable. However, if exceptions quantitatively increase to a certain level, they can lead to an adjustment of the theory system (chap. 2, sec. 3). "When two ideas conflict with each other, all the forces of demagogy are activated. And it is almost always a third idea that emerges triumphant: one woven from exoteric, alien-collective, and controversial strands" (chap. 4, sec. 4).

Differences of opinion and conflict between thought styles can lead to a more fundamental transformation. Great transformations in thought style, Fleck suggests, "often occur during periods of general social confusion. Such periods of unrest reveal the rivalry between opinions. . .[and] a new thought style arises from such a situation" (chap. 4, sec. 4).

Discovery in science, whether modification or transformation, and whether of a theory or its thought style, is a complex, socially conditioned product of collective effort. An individual belonging to more than one collective at the same time may be led to create "a special style on the borderline of the field" of research (chap. 4, sec. 3). By standing at the intersection of several thought collectives, the creative scientist can form the nucleus of a new thought style.

Fleck claims that it is the democratic duty of every individual scientist to recede into the background. "All research workers are, as a matter of principle, regarded as possessing equal rights. Each must equally withdraw his own individuality into the shadow, as it were, in the service of the common ideal" (chap. 4, sec. 5). Yet scientists typically engage in competition, and priority disputes are not infrequent, which would suggest that Fleck's democracy of science remains incomplete.

Critical Reception

Reviews of Fleck's monograph appeared shortly after its publication and mainly in medical journals (see list at end of Bibliography below). Those that we have in hand take a positive stance on the book and generally rate it very high. Fleck is praised for his breadth and depth of evidence and his understanding of the problems involved. All the reviewers recommend the monograph, with varying enthusiasm, and some consider it a valuable model for further research.

A few of the reviews provide a critical and comparative analysis. Fischer agrees with Fleck on the cultural conditioning of thought style but suggests that he offers nothing beyond that already indicated by Max Weber, Max Scheler, and Karl Mannheim on the sociology of concept formation. What *is* new is the manner in which Fleck applies these principles to *scientific* cognition. Particularly important is the way in which scientific findings arise from nonobjective factors. Fischer doubts that cognition should be designated the activity of man that is most strongly conditioned socially. He concedes that Fleck is justified in rejecting the individualistic tendency in psychology (chap. 4, sec. 4) but feels he has gone to the other extreme by reducing the individual to the collective. Petersen is pleasantly surprised that Fleck follows the "new German thought style," which denies the absolute and unconditioned character of science and always considers it an integral part of the total culture. In applying the principles of sociology and culture to science, Fleck is building upon a foundation already laid by Viktor von Weizsäcker and others. Petersen feels that Fleck overstressed the constraining role of the thought style and underrates individual creativity. He criticizes Fleck's preference for the Russian "collective" to the German "community," for in a collective all the individuals are exchangeable, but the personalities of a community are not. Fleck has correctly observed that the contribution of Schaudinn was not his alone but that of a group (chap. 2, sec. 4). But whereas this group would be lost in anonymity if treated as a collective, the specific personalities of members of any group would be preserved if analyzed on the basis of community. It is not true, Petersen contends, that just any group would do. Even now, there is perhaps no aspect of Fleck's theory more potentially controversial than the perennial issue of "collective" versus "community," fraught as it is with implications concerning the status of individual personalities. Petersen does not concur with Fleck's characterization of vademecum science as impersonal, though he agrees with Sigerist's suggestion that a handbook can be considered the burial ground of science. By tending to identify a fact with knowledge about a fact or with an interpretation of a fact, according to Petersen, Fleck is adopting a position of extreme idealism. He may have been led to this position, Petersen suggests, by the complex and nebulous character of the particular fields of medicine with which he is concerned. His analysis may therefore be more appropriate to the history of medicine than to other areas of science. This latter point is reinforced by Lauriers, who feels too that Fleck has not properly taken individual differences into consideration. Kroh notes that it would be of interest to investigate the function of racial factors in the origin of thought styles. Bing urges every physician and scientific researcher carefully to examine Fleck's work, although another reviewer (anonymous, in *Natur und Kultur*) sadly acknowledges that this rich treasure of the first rank concerning medical philosophy may not yet be fully appreciated.

Conclusion

Fleck has provided a theory of knowledge whereby tentative pre-ideas can develop into scientific facts by means of a process of collectivization. Specifically, he has drawn attention to the discovery of the Wassermann reaction as the fulfillment of the ancient belief that a connection exists between syphilis and the blood.[2] The pre-idea of impure blood in syphilitics was present in several thought collectives. One considered syphilis to be an ethical issue and called it "carnal scourge" (*Lustseuche*); certain physicians linked syphilis with other skin afflictions; while another collective linked the symptoms to mercury poisoning. Each in its own way led to the insistent call for blood tests. The etiological or causative approach to syphilis led to bacterial findings which allowed the ancient pre-idea to become established as a scientific fact.

Fleck was convinced of his fundamental position: that scientific knowledge is culturally conditioned and that to understand its nature requires an epistemology sociological and comparative in character. He expected that a bond of solidarity would be established between his presentation and the intellectual mood of the reader (chap. 4, sec. 5). Except for a few rather favorable reviews, however, no such solidarity obtained for several decades. Thomas S. Kuhn discovered in Fleck points of common interest and during the 1960s rode the crest of a great wave of interest. In the late 1970s, over forty years after its initial publication, Fleck's monograph has finally come of age. To use his own term, there has been a shift in thought style in favor of Fleck. If this does not justify his thesis,[3] it certainly adds to its timely interest.[4]

Notes

Where bibliographic references are given by author and date only, the reader may turn to the Bibliography below for titles and publication details.

Foreword

1. A more extended account of the "revelation" is included in the preface to a recent selection from my essays, *The Essential Tension: Selected Studies in Scientific Tradition and Change* (Chicago, 1977).

2. R. K. Merton, *Science, Technology, and Society in Seventeenth Century England* (New York, 1970), p. 221 n. Merton's monograph appeared originally in 1938.

3. Hans Reichenbach, *Experience and Prediction* (Chicago, 1938), p. 224 n.

4. Ibid.

5. Thomas S. Kuhn, *The Structure of Scientific Revolutions* (1962; 2d ed., enlarged, Chicago, 1970), pp. vi–vii.

6. In quoting this and the following fragments from Fleck's texts, I have added and subtracted italics at will. Readers may wish to check the translation, below.

7. Wittgenstein's *On Certainty* (Oxford, 1969) is particularly relevant.

Preface

1. Maurice Crosland, ed., *The Emergence of Science in Western Europe* (London, 1975).

2. Karl Mannheim, "Das Problem einer Soziologie des Wissens," *Archiv für Sozialwissenschaft und Sozialpolitik* 53 (1925): 577-652; compare, for example, Rudolf Carnap, *Der logische Aufbau der Welt* (Berlin, 1928), preface; Hermann Friedmann, *Epilegomena: Zur Diagnose des Wissenschafts-Zeitalters* (Munich, 1954), p. 18; Herbert Sultan, "Gesellschaftliche Strukturwandlungen und nationalökonomische Theorie," *Zeitschrift für die gesamte Staatswissenschaft* 109 (1953): 602-14 (being an application of *Denkstil* to economics); and the volume by H. Schüling entitled *Denkstil* (Ratingen, 1967). The Brockhaus Encyclopedia, 17th ed. (1968), has an entry under *Denkstil;* and, more recently, the *Frankfurter Allgemeine Zeitung,* 9 November 1974, carried an interesting review of a German translation of George Steiner, *Extraterritorial* (London, 1972; Frankfurt, 1974) under the caption "Kulturjournalismus als Denkstil." In their translation of Mannheim's *Ideologie und Utopie* (Bonn, 1929) Louis Wirth and Edward Shils of the University of Chicago rendered *Denkstil* sometimes as "style of thinking" or as "style of thought" but typically as "thought-style"; compare *Ideology and Utopia: An Introduction to the Sociology of Knowledge* (London, 1936) p. 136, with the fourth German edition (Frankfurt, 1965) p. 133, both of these on pp. 30, 36, 38, 45, 46, and passim.

3. L. Fleck, "Zur Frage der Grundlagen der medizinischen Erkenntnis," *Klinische Wochenschrift* 14 (1935): 1255-59 (citation, p. 1259). In this important précis of his book, Fleck develops the further point that medical science, as an important facet of the development of modern civilization, can and should complement natural science in approaching rational epistemology and historiography of science.

4. Thaddeus J. Trenn, "The Non-Rational Dimension of Natural Science," *Deutsche Vierteljahrsschrift für Literaturwissenschaft und Geistesgeschichte* 50 (1976): 1-13. I presented and distributed copies of a short paper on the Fleck project at the Philadelphia meeting of the History of Science Society, 30 December 1976, and at a public lecture held at Harvard University, 18 January 1977, where I also displayed the Harvard copy of Fleck's book, complete with the library stamp, indicating that it was borrowed in late 1949 and early 1950—presumably by Kuhn.

5. W. Baldamus, "Ludwig Fleck and the Development of the Sociology of Science," in *Human Figurations: Essays for Norbert Elias,* edited by P. R. Gleichmann, J. Goudsblom, and H. Korte (Amsterdam, 1977).

6. Among the current literature references on Fleck the Marxist approach to Fleck of Wittich in *Deutsche Zeitschrift für Philosophie* 26 (1978): 105-13 and 785-97 might be mentioned. A second edition of the German original is planned by Benno Schwabe & Co., Basel, Switzerland.

One   How the Modern Concept of Syphilis Originated

1. Bloch 1901–11, 1:138. Baas 1876, p. 259. Hergt 1826, pp. 47, 56.
2. Bloch 1901–11, 1:26.
3. Rinius, p. 18.
4. Brassavolus, *De morbo Gallico,* cited from Bloch 1901–11, 1:17.
5. The name *morbus venereus* was allegedly introduced by Bethencourt (1527), but the venereal nature of the disease had already been stressed before. Widman, 1497: "It is therefore essential to refrain from sexual intercourse with a woman afflicted with boils, indeed even with a healthy one with whom a man afflicted with boils had had intercourse a short while ago, to avoid the risk of contagion" (Geigel 1867, p. 11). Almenar, 1502: "Men should refrain from close contact with infected persons and especially from sexual intercourse with an infected woman, for this disease is contagious" (ibid.).
6. One of the old names of syphilis. Sudhoff's views on the diagnoses of syphilis as early as the fourteenth century are not generally accepted. Discussion of the disease does not, at any rate, begin before the end of the fifteenth century. Sudhoff 1913, pp. 13–14. [Cf. Lesky 1959, passim.— Eds.]
7. Frizimelica, p. 33.
8. Ibid.
9. Borgarutius, p. 178.
10. Hergt 1826. Characteristically, the book bears the following dedication: "To the mercuric and mercurous oxides and salts the author pays his deepest respects for their great services to suffering humanity."
11. Tomitanus, p. 66. Tomitanus claims to have established counter-evidence with the following case history: "A young man of twenty-two, of lively constitution," was a conscientious and chaste student at Padua. Bad company introduced him to a "winsome wench." "On the following day his foreskin began to hurt, but he thought nothing of it. The following day the pain increased, and when he finally inspected it, he saw some slight erosion in another part of the glans with reddening where rot had set in. After a fortnight a boil occurred on the thigh, which the physician lanced and cleaned out. After three months he began to feel pain in all joints and to lose his hair, and became disfigured, thin, bluish, listless, moved with difficulty, was depressed, sighing, and incapable of any action. . . . About mid-spring he took the Guaiacum prepared on his physician's advice, and was completely cured and discharged." Tomitanus demands from the doubters that "the proponents should say by way of contradiction whether this disease, arising from this cause, is one of the old afflictions, or a new

and unheard-of one?" If we read this story in the naive hope that here we have "holy observation," "simple, decisive seeing," we shall soon be disappointed. The case, is at any rate, not unambiguous. The incubation period of syphilis is never twenty-four hours, although this may be so with soft chancre. Purulent boils are also symptoms of soft chancre, never of syphilis. On the other hand, the secondary symptoms described, which occurred after three months, are not those of soft chancre. These might indicate syphilis, but equally likely some other, "nonspecific," disease. The Guaiacum extract—a popular contemporary cure for syphilis—indicates merely that the physician suspected syphilis. The alleged curative effect proves nothing, because Guaiacum extract is not a specific remedy for what we today designate syphilis.

This whole case history represents a schematic, vague image of the mythical carnal scourge (chastity, seduction, punishment inflicted on the genitals, general malaise, cure by Guaiacum). It cannot be translated into modern medical language. To us it is by no means a "straightforward case." A combined infection of soft chancre and syphilis or of soft chancre and an independent, nonvenereal disease could also after three months display similar symptoms and a similar sequence. (See also chap. 4, sec. 3, n. 5.)

12. Hermann 1891. He had followers, and such views could be mentioned and quoted even twelve years after Neisser discovered the causative agent of gonorrhea in 1879 and two years after Ducrey discovered the agent causing soft chancre in 1889. Hermann based his theory on the fact that he found mercury in the secretions of patients diagnosed by others as suffering from constitutional syphilis, just as with workers in mirror factories who suffered from mercury poisoning and also exhibited various general and sometimes very similar symptoms. Hermann saw in this mercurialism a constitutional, hereditary, and very polymorphous disease. He treated his patients without Hg and claimed never to have observed relapses, but only fresh, and possibly recurring, infections. His views therefore in no way constitute a simple error but are a self-contained system of opinions; specifically the realization of the postulate "Back to the pre-Hg era."

13. To Simon (about 1850) the "so-called modern great pox [as carnal scourge] is nothing but a special variant of the very ancient leprosy, which at the close of the fifteenth century raged uncontrolled owing to exceptional circumstances." Simon 1851–52, p. 3.

14. Hergt 1826, p. 78. These physicians lived during the eighteenth century.

15. John Hunter, 1728–93.

16. Philippe Ricord, 1800–1889. [The unitarians were so designated because they advocated a unitary system.—Eds.]

17. In two versions: a French and a German theory of duality.

18. Conceptions seen with the wisdom of hindsight often appear to be economical, especially when we have become used to them. An existing establishment is always more economical than a planned new one if there is no promise of a return on the investment within a specified period from the profits of the inherently more economical new establishment. Opinions are of limited duration anyway, so costly modifications of them are almost always uneconomical. I doubt whether economy of thought has ever produced any practical decisions, except perhaps in cases involving very minor and insignificant problems.

19. We read, for instance, in Thomas Sydenham, 1735, p. 3: "But especially the fevers...names by which they are distinguished through some change in the blood."

20. Montagnana, p. 3.

21. Ibid.

22. Tomitanus, p. 74.

23. Ibid., p. 88.

24. Ibid., p. 113.

25. Cited from Geigel 1867, p. 12.

26. Geigel 1867, p. 39.

27. Ibid., p. 70.

28. Bloch 1901–11, p. 98.

29. Bruck 1924, p. 1.

30. Wendt 1827, p. 9.

31. Bierkowski 1833, p. 36.

32. Hergt 1826, p. 58.

33. Cited from Geigel 1867, p. 7.

34. Cited from Geigel 1867, p. 19.

35. Geigel 1867, p. 223. Here is also a more detailed attempt to analyze the changes in the blood.

36. Reich 1894. According to Reich, syphilis also includes caries of the bones in general; psoas and lumbar abscesses; any form of phthisis in any age group, tuberculous diseases wherever they may occur, rickets, diseased nerves, diseased souls, sickly constitutions, etc.

37. Waller's experiment "is said to have succeeded in 1850." Hermann 1891, p. 24. Is *said* to have! The author has his doubts, because the results are not compatible with his theories. There were several other experimenters besides Waller; Anonymous from the Palatinate [Dr. J. Bettinger, according to Sudhoff 1922, p. 450.—Eds.], Lindwurm, Pellizzari, and others.

38. Hermann 1891, p. 26. Hermann here sees transmissions only through the syphilitic secretions of the skin, not through the blood.

39. Ibid., p. 32.

40. Bruck 1924, p. 1.

41. Schuberg and Schlossberger 1930, p. 582.

42. Levaditi's neurotropic virus is, however, often construed as *Spirochaeta cuniculi.*

43. In practice, not always even then, owing to frequent failures of experimental cultures and inoculations.

44. According to Ermoljewa, harmless water vibriones cannot with any certainty be distinguished from cholera vibriones. Lehmann and Neumann 1926–27, p. 540, state: "When the cholera vibrione was discovered, its properties seemed so characteristic that it was expected to be easily distinguishable from other [types of] bacterium. Since then, first a few, then more and more, and ultimately such vast numbers of vibriones have been found in the human environment, that they have long ceased to be given separate names."

45. For instance, the relation of syphilis to *Frambösia tropica,* as well as the so-called rabbit spirochaete, is still disputed.

Two   Epistemological Conclusions

1. At first sight one might think that this claim referred only to abstract concepts. Diseases as such are thought not really to exist but only diseased persons. Syphilis is construed as a state that sick people are in, rather than as a concrete concept. Against this I would argue that no clear-cut distinction can be made between concrete and abstract elements. The entire classification is based upon a very primitive way of thought. We shall nevertheless investigate later, in the light of this classification, so-called direct experience, which is considered the most concrete of all.

2. Kirchberger 1922 and Lange 1905.

3. Lange 1905, p. 37.

4. Metzger 1929, on Hornbostel's work. ["Ideophones," here used to render the German *Lautgebilde,* according to the *Oxford English Dictionary* was introduced by A. J. Ellis to designate a sound or group of sounds denoting an idea.—Eds.]

5. Mach 1908, pp. 521 ff.

6. Kant 1921–22, vol. 3, p. 22.

7. Wundt 1893–95, vol. 1, p. 446.

8. Ehrlich was referring to his work on toxin analysis.

9. This was Nägeli's undoing in his controversy with Cohn and Koch. [Cf. Philip Hadley, "Microbic Dissociation: The Instability of Bacterial

Species with Special Reference to Active Dissociation and Transmissible Autolysis," *Journal of Infectious Diseases* 40 (1927): 8.—Eds.]

10. Paracelsus, *Von den unsichtbaren Krankheiten und ihren Ursachen* (On the causes of invisible disease), in the transcribed edition by Richard Koch and Eugen Rosenstock, p. 21.

11. The original edition of Huser (1589) reads: "Ist das nit ein wunderbarlich Werck durch Gott, das d' Mensch soll lebendig auff Erden ein Teufel zuhaben, erscheinen?" [The Koch-Rosenstock edition used by Fleck reads "Ist es nicht eine wunderbare Tat Gottes, dass der Mensch lebendig auf Erden einen Teufel zu haben scheint."—Eds.]

12. Cf. Mach 1908, p. 494.

13. Complete with names, precise figures, and repeated measurements.

14. The same opinion is held by other authors. Cf. Bartholin 1673.

15. See figure 1.

16. About 1520 (according to Roth 1892, p. 41).

17. Even today speculative epistemology is taught as a science, in which its speculative investigations are almost always limited to a few symbolic examples and logical connections, preferred over and above all other connections, between the objects of investigation.

18. "In purely dissecting anatomy, a marble statue with its delightful, simple art form would be just hacked into a rubble heap of marble fragments" (Bölsche 1907, p. 140).

19. Anyone wanting to convince himself of this impossibility should read the controversy between Bethe and the anatomists in *Klinische Wochenschrift* 1928. [The other protagonists were Ackermann, Fick, Fröhlich, Göppert, Goldstein, and Petersen.—Eds.]

20. Nägeli 1927, pp. 50–51. In spite of the accusation implied by Nägeli, I do not believe in a simple case of bad faith on the part of Kammerer, who was an original and diligent research worker.

21. "Consolidation" [*Verdichtung*], as Jerusalem calls it.

22. Although nobody would refuse to attribute the creation of intellectual products such as language, folksongs, folklore, and others to a collective.

23. According to Jerusalem, from his preface to the German edition of Lévy-Bruhl 1926, p. vii.

24. Lévy-Bruhl 1926, p. 1.

25. Ibid., p. 2.

26. Ibid., p. 5.

27. Ibid., p. 10.

28. Ibid., p. 11.

29. Gumplowicz 1905, p. 268, quoted according to Jerusalem 1924, p. 182.

30. Jerusalem 1924, p. 183.

31. Ibid., p. 188.

32. Ibid., pp. 191–92. [Jerusalem had introduced this idea of *soziale Verdichtung* as early as 1909.—Eds.]

33. Ibid., p. 192.

34. Lévy-Bruhl 1926, p. 336.

35. Ibid., p. 342.

36. Ibid., p. 337.

37. Ibid., p. 339.

38. Jerusalem 1924, p. 188.

39. Ibid., p. 193.

40. Ibid. But we soon read: "Not every observation by an individual must in itself be valued as an experience. Only after a stock of general and well-confirmed knowledge has formed as a result of mutual agreement and reinforcement in the course of continued cooperation of the intellects involved should we speak of experience. General and well-confirmed experience, however, must be considered the sole criterion of truth" (p. 199). Confrontation of these contradictions does not constitute a criticism of Jerusalem. It merely exemplifies that when new thought styles are evolving, contradiction sets in as an expression of the intellectual "contest of the fields of view."

41. Nor can assent be given to Jerusalem's view about the origin of logic. "The origin of logic is closely connected with the evolution of the idea of the whole of mankind as a great entity. General logic constitutes the proportion of logical supra- and subordination, which is valid for all human intelligence. The further development of this leads to increasingly comprehensive generalization, in which general and well-confirmed experience is defined, economically ordered, and formulated more and more precisely" (ibid., p. 206). This is much too schematic. Are the primitive races also part of mankind construed as an entity or not? They have a different logic, which is no more valid for the whole of mankind than ours is. And how are the mystics and gnostics living among us to be classified? The concept of a thought collective comprising the whole of *Homo sapiens* is of little use, because the intellectual interactions between different types of human society are too weak.

Three   **The Wassermann Reaction and Its Discovery**

1. "If the modern biologist wants to form an objective picture of the living world, he must rid himself of all ideas based on a subjective approach. It is sometimes not at all easy to get rid of such prejudices completely. Man's own consciousness of himself as a self-contained whole

or entity arouses in one the instinctive notion that the whole living world is divided into a certain number of such units which we call organisms." Gradmann 1930, p. 641.

2. Ibid., p. 666.

3. Cf. Hirszfeld 1931, p. 2153.

4. See figure 2.

5. Bruck 1924, p. 3.

6. Weil 1921, p. 967.

7. Wassermann 1921, p. 1195. According to Bruck, a member of Wassermann's team, the stimulation came from Neisser, not from Althoff. [Bruck 1921, p. 581.—Eds.]

8. There may have been occasions when Wassermann and Bruck exchanged roles. [Bruck later claimed sole credit for the *experimental* work which, by luck alone, turned the idea of Wassermann and Neisser into a useful method; Bruck 1921, p. 581.—Eds.]

9. Wassermann, Neisser, and Bruck 1906, p. 745.

10. Ibid., p. 746.

11. So designated according to the reaction scheme, but it is doubtful whether these are so-called true amboceptors.

12. My italics.

13. Bruck 1924, p. 3.

14. Wassermaan, Neisser, Bruck, and Schucht 1906.

15. Weil (1921, p. 967) wrote about these experiments: "The study of this paper could not fail to convince everybody firmly that here was a reaction which functioned with wonderful precision, especially in the demonstration of spirochaete antigen. In 69 extracts from syphilitic tissue, the specific antigen was shown 64 times, whereas 7 extracts from the brains of progressive paralytics all reacted negatively (at the time, proof of the presence of spirochaetes in the brain of progressive paralytics had not yet been established). Fourteen control tests on nonsyphilitic tissue were uniformly negative."

On antigen detection, which later was completely discarded, Bruck wrote in 1924: "To test whether complement fixation in the context of antigen detection might *also* become a clinically usable technique, Neisser, Bruck, and Schucht tried to test serum containing syphilitic antibodies with blood extracts from syphilitic patients. These results at first appeared promising but were not reliable enough to allow safe conclusions to be drawn from them. It seems to be evident, however, that even blood extracts from syphilitic patients differ biologically from those of healthy individuals. Whether this difference is based on an increased antigen content in syphilitic blood or on other factors is an open question. Since we now have at least some information about the true nature of the reaction, these experiments are of only secondary importance" (Bruck 1924, p. 7). The word "also" stressed by me in the first sentence does not correspond to the situation as it was in 1906, but to the "secondary importance" these experiments appear to have had for Bruck in 1924.

16. Citron 1910, p. 187.

17. Weil 1921, p. 968.

18. This excellent cholesterol addition is the result of a certain confusion among serologists over concepts. No albumins are present in alcoholic extracts; the serologists therefore looked for the active principle in the group of alcohol-soluble lipoids, whose chemistry, however, they did not quite understand. The addition of cholesterol is probably the cause of an increase in the instability of the colloidal state of the extract.

19. Bruck 1921, p. 581.

20. Weil 1921, p. 969.

21. Laubenheimer, in Kolle, Kraus, and Uhlenhuth 1930, vol. 7, p. 217. Laubenheimer too fails to see how the Wassermann reaction changed, and does not understand that we must speak not of proving its worth but of development.

22. Plaut 1931, p. 1463.

23. Wassermann 1921, p. 193.

24. I wish to emphasize strongly that it is not my intention either to minimize the merits of a research scientist or even to discuss merits. I have listed various views on authorship and contribution to the discovery of this extremely important reaction only for epistemological purposes: to show that everybody makes mistakes. As for reverence for a master, it is not success that is the proof of greatness but the nature of his effort. I do not believe that a revered research worker becomes greater when he is set up on a pedestal rather than portrayed as a living human being.

25. Let me draw attention once again to the unique ethical emphasis in discussions of this disease. We find, for instance, in the previously quoted brochure by Reich the following description of an allegedly syphilitic family: "The entire family Cattolupino was beyond all control, exhibiting excesses, violence, coarse behavior, arrogance, exaggerated self-esteem, mistrust to the extreme limits of what is possible, contentiousness, contrariness, argumentativeness, virtuosity in babbling, fault finding and knowing all the answers, no knowledge of human nature, profound ignorance, complete absence of either tact, consideration, or caution, contemptible servility to mammon," etc. "All these moral evils must necessarily be based on physical evils of a serious nature." "It must be accepted with certainty that acquired syphilis turned into inherited syphilis and was poorly treated." There is no other disease that has ever to this extent been

regarded as the cause of moral decay. Although leprosy also had strong emotive connotations, these pointed to fate and not to morals. Only syphilis, associated with sexuality, had a moralistic tinge. Even today a wide sector of the public associate sexual relations with morals. Geigel expressly spoke of the "specific malicious trait" with which syphilis was invested from the very beginning through its connection with sexual intercourse (Geigel 1867, p. 4). "Through its regeneration so darkly connected with the secretive act that is responsible for the propagation of the human race, ever since its appearance at the end of the fifteenth century, which was outrageous for public and doctors alike, it has been like a nightmare for the most delicate relations, clinging like the breath of pestilence to youth and beauty, fastening like an evergrowing and monstrous burden of sin onto a single lapse as well as poisoning the blood of unborn innocent children" (ibid., p. 1).

26. See above on the repercussion of the Wassermann reaction upon the concept of syphilis.

27. Bruck 1924, p. 4.

28. Laubenheimer, in Kolle, Kraus, and Uhlenhuth 1930, vol. 7, p. 216.

Four   **Epistemological Considerations**

1. This is the "ultramethodics with a personal undertone" that Meier demanded from the first critics of the Wassermann reaction which precisely underlines the personal factor in the emerging truth. [Cf. Weil 1921, p. 968.—Eds.]

2. Fleck and Elster 1932.

3. Carnap's system (Carnap 1928) will perhaps be the last serious attempt to construct the "universe" from "given" features and from "direct experience" construed as the ultimate elements. Criticism is unnecessary, since Carnap himself has already gradually relinquished this point of view (Carnap 1931). Concerning his viewpoint, according to which the absolute character of protocol statements is already rejected (Carnap 1932), one would hope that eventually he might discover the social conditioning of thought. This would liberate him from any absolutism in the standards of thought, but of course he would also have to renounce the concept of "unified science."

4. For the sociology of science it is important to state that great transformations in thought style, that is, important discoveries, often occur during periods of general social confusion. Such "periods of unrest" reveal the rivalry between opinions, differences between points of view, contradictions, lack of clarity, and the inability directly to perceive a form or

meaning. A new thought style arises from such a situation. Compare the significance of the Early Renaissance or of the period following the Great World War.

5. If we look at a fact torn from its developmental context, it is in the nature of such isolation that only one definite inevitable connection within this artificially isolated relation will appear important. The active parts appear as disturbances to be fought. They play the role of friction, only the retardation of which, not the advance, is felt during motion. But if facts are seen in their context and development, it will soon be realized that the roles of the active and passive components of any knowledge are often interchanged. Also, what contributes actively and what passively to the act of cognition depends on the properties of a thought style. As for the fact of the relation between the Wassermann reaction and syphilis, the concept of "syphilis" is for us today the more actively constructed one, that of the "Wassermann reaction" appears to be somewhat less so, and the kind of relation must as such be considered passive. To one who regards this affliction simply as the carnal scourge meted out by a superior being as punishment for sinful lust, syphilis is a passive (objective) given. But the Wassermann reaction appears as an active (artificial) construction, unrelated to syphilis because, after all, gonorrhea and other "venereal diseases" must be part of syphilis as he conceives it. Even if he limited syphilis in the same way as we do today, he still could not see its relation to the Wassermann reaction, had he in some way been introduced to it, because the result of the reaction depends on the stage of the disease. It is therefore incompatible with the concept of an absolute disease with an objectively independent existence. Thinking logically, he would declare not only the reaction but also the limitation on syphilis, and the syphilis classification, all to be a piece of godless casuistry, because the carnal scourge can be diagnosed better by a moralistic analysis of the patient's conscience than by a laboratory analysis of his blood. In the sixteenth-century case history in chapter 1, note 11, the diagnosis was based partly on the anamnestically established sin of the previously chaste and pious young man. This is not a case of syphilis, according to contemporary thinking, but one of carnal scourge as understood at the time. To the author of this case history, the Wassermann reaction and its relation to syphilis would have been completely obscure and a mistaken finding. In this piece of sixteenth-century casuistry one relation is of particular interest. The case as outlined corresponds to the general aspect of syphilis in the current sense of the term, but the details given militate strongly against it. The sequence of symptoms corresponds to syphilis: primary infection after the sex act, followed by affection of the glans and, after a certain interval, general symptoms. But the details about time intervals and

individual symptoms as described in chapter 1, note 11, do not agree with
the modern aspect of this disease entity. The form was already vaguely
outlined, but the details were to be filled in only much later. The form was
"searching," so to speak, for its realization. It was tentatively applied to
cases in which it was untenable. Was this simply error? Or does the birth
of a comprehensive concept always appear as a contour which initially is
too broad and too vague and only later acquires more substance and thus a
narrower profile?

6. Uexküll (1928) to some extent sees the problem of the subjective
conditioning of the world view in a similar light: "The physicists, with
their faith in the absolute existence of an objective world, have reached a
dead end" (p. 30). "Opposed to this the biologist claims that there are as
many worlds as there are subjects, that all worlds are worlds of appearance
that can be understood only in connection with the specific subject" (p.
61). But on p. 231 he says: "The universe consists of subjects, each with
its own environment, which are integrated by means of groups bound by
common function into an organized entity." *So there is* then a universe
and an organized entity which is not subjective? Apart from many unac-
ceptable statements in Uexküll's theory of environments, his solution lacks
a correct assessment of the social factor in knowledge. Nor can we agree with
the division into a world of perception and a world of action, because it is
practically untenable and leads only to useless metaphysics. Does a "perception" occur only passively, without any "action" and vice versa? Can
an "action" and its effect be assessed other than by a "perception"?

7. The effect of social communication on works of the mind is well
known to sociologists. Georg Simmel, for instance, in *Soziologie* (1908),
especially in chapter 2, discusses social differentiation. The well-known
book by Le Bon, *Psychologie des foules,* deals almost exclusively with the
momentary mass, mainly in its violent state of emotion. Le Bon seems to
be far removed from the psychology of the disciplined, quietly working
community, which may not even be concentrated in a single locality. He
describes in his book, among other situations, a case of mass suggestion
which induced the entire crew of a ship, while searching for a boat in
distress, to see this craft and its crew and even to hear shouts and see
signals. This collective hallucination was suddenly dispelled but only during
the last minute of approach. The "boat" turned out to be a tree with
branches and leaves drifting in the water. This case could be considered
the very paradigm of many discoveries. The mood-conforming gestalt-
seeing and its sudden reversal: the different gestalt-seeing. Suddenly one
can no longer even understand how the previous form was possible and
how features contradicting it could have gone unnoticed. The same situation obtains in scientific discovery, only translated from excitement and

feverish activity into equanimity and permanence. The disciplined and even-tempered mood, persisting through many generations of a collective, produces the "real image" in exactly the same way as the feverish mood produces a hallucination. In both cases switch of mood (switch of thought style) and switch of image proceed in parallel. Because Le Bon knows only the excited momentary mass, he sees in any socialization merely a degradation of psychological qualities. McDougall, in *The Group Mind*, attempted to save the undeniable positive values of socialization by according the power of "equipping the mass with the attributes of the individual" to the organization. Freud, in *Massenpsychologie und Ich-Analyse*, seeks to resolve the community of action and feeling within a mass into individual psychological elements by assuming that the mass-individuals are identical with each other and that the leader is a common ego ideal. Neither of these authors can explain the specific, nonadditive element of the mass psyche. Hans Kelsen (1922) sharply attacks the concept of the mass psyche. "Because psychological events are possible only in individuals, that is, in the psyche of separate persons, anything [allegedly] superindividual, beyond the individual psyche must have a [suitable] metaphysical character." "We might assume that besides the individual psyche there is also a collective psyche filling the space between the individuals and embracing all individuals. But if logically thought through, and because experience shows that a psyche without body cannot exist, this concept must lead to the assumption of a collective body, similarly distinct from the individual bodies, in which the collective psyche is supposed to be lodged" (p. 125). "From the point of view of epistemological critique, this mythological method is seen as that tendency—to be suppressed because it is mistaken—to reinterpret as concrete objects the relations both to be determined and determinable by cognition alone, as well as to reinterpret function as substance" (p. 138). This description is perhaps deliberately exaggerated for reasons of polemics. Who would seriously imagine a space-filling psyche? Besides, the hypostatization or reification of society (p. 139) which Kelsen fears so much is really quite harmless, because this substance would be functionalized at the same time. In recent science there is no concept of substance in the same sense as it was known only about fifty years ago. Not much more remains of this concept than an almost illusory peculiarity of thought to be explained psychologically and in terms of the history of ideas. How does one arrive at the structure [*Gebilde*] of "body" as a specific form to be directly perceived? There is no doubt that in everyday life, with the several senses of feeling, pain, muscles, vision, we actually perceive "bodies" without any difficulty and without, drawing conclusions or making conventions. But upon analysis these "bodies" dissolve into functions.

If workers in highly exact sciences such as physics do not shrink from making use of statistical data, such as average values or probability values, which correspond not to any "actual" appearance but to a hypostatized fiction, and indeed consider an "actual" appearance much less "genuine" than this fiction, we shall probably have no reason to fear any damage caused by the introduction of the concept of thought collective. If, as I hope, it benefits knowledge, it will be legitimized. I consider objections on principle such as those mentioned here altogether obsolete, because philosophical principles are like money. They are very good servants but very bad masters. Principles should be made use of, but not blindly accepted as guides.

The boundary line between that which is thought and that which is taken to exist is too narrowly drawn. Thinking must be accorded a certain power to create objects, and objects must be construed as originating in thinking; but, of course, only if it is the style-permeated thinking of a collective.

[For a further, independent discussion of these issues, see Alexander Gerschenkron, "Figures of Speech in Social Sciences," *Proceedings of the American Philosophical Society* 118 (1974): 439–40, note 40.—Eds.]

8. Gottstein 1929, p. 30.

9. It is characteristic of a closed community to claim to be *tout le monde,* all the world, everyone—to look down on outsiders or simply to declare them to be nonexistent.

10. Simon 1851–52, p. 15.

11. There is a neglect of morphology, of the biology characteristic of bacteria, and a preference for examining pure cultures, coupled with neglect of population research, errors of taxonomy, and so on.

12. Ramsay 1913, p. 191 [1908, p. 148].

13. Ostwald 1906, pp. 25–26.

14. Ramsay 1913, p. 58 [1908, pp. 21–22].

15. Paracelsus, *Von den unsichtbaren Krankheiten und ihren Ursachen,* Huser's edition, vol. 1, p. 247. The transcription of Koch and Rosenstock reads "Wenn ihr einen Glauben wie ein Senfkorn habt und seid schon irdische Geister, wieviel mehr wird euch werden, wenn euer Glaube ist wie die Melonen? Wie sehr werden wir die Geister übertreffen, wenn er ist wie die grossen Kürbisse und so weiter." [The Huser edition used in the text reads "So ihr habend einen Glauben als ein Senfkorn und seind Irdische Geiste, wie vil mehr wirt es euch werden, wenn ewer glauben ist wie die Melonen: wie hoch werden wir die Geiste übertreffen, wenn er ist wie die grossen Cucurbiten."—Eds.]

16. Paracelsus, *Von der Gebärung der empfindlichen Dinge in der Vernunft,* Huser's edition, vol. 1, p. 350.

17. Schreger 1755.

18. Compare this with the previous quotation from Schreger. This "error" regarding the concept of weight [*die Schwere*] recurs through several centuries and still survives in popular lore.

19. Similarly, there are motifs common to the concept of syphilis in modern science and to that of a few centuries ago. From this point of view, the historical development of every scientific concept could be investigated according to a theory of thought style.

20. In his theory of "signatures," Paracelsus von Hohenheim considered that the appearance of an object indicated what it cured. Thus the alleged healing powers of eyebright could be recognized from the fact that its flower displayed the pattern of a human eye (Koch and Rosenstock 1923, p. 24). [Compare and contrast their comment with Paracelsus, Huser edition, vol. 1, p. 246, at *Eufragia,* and vol. 6, p. 357, at *Augentrost,* in book 9 of Paracelsus' *On the Signatures of Natural Things*—Eds.] The testicle-shaped orchis root is supposed to be an effective remedy for diseases of the testicles (Baas 1876, p. 316). Very similar claims can be found in ancient Indian medicine (yellow plants for jaundice) and in the folk medicine of some Western nations.

21. Møller and Müller 1914. This compendium, written for general practitioners, has about 510 pages, compared with the 850 pages of Bartholin's *Anatomy.* The printed area of the page is almost the same in each case.

22. Fontanus 1642.

23. See figure 3. There is a special compulsive mental association between the skeleton and death. Fontanus 1642, p. 3: "In my view, when ghosts or nocturnal shadows frighten a person, they do it in the form of skeletons." [Cf. Kuhn's Foreword, p. viii, above.—Eds.]

24. See figure 4.

25. See figure 5.

26. Toldt, for instance, confines himself to: "Sesamoid or articulated bones are osseous growths, mostly small inclusions in tendons."

27. See figure 3.

28. This *postulate concerning a maximum of information* must be separately stressed, because it is an outstanding characteristic of the thought style of modern natural science. It can be formulated as follows: "No system of knowledge [*eines Wissens*]—for example, about a chemical compound or a biological species—must be regarded as closed in such a fashion that possible new findings might be rejected as superfluous." To assess the difference, compare the diametrically opposed position of a dogmatic knowledge that is regarded as completed. That is also a democratic feature

of the thought style of natural science, which does not accord any preference or privileged position to previous over new knowledge.

Biographical Sketch

1. The story of Fleck's work during the war is also described in Stefan Szende, *Der letzte Jude aus Polen* (Zurich and New York: Europa Press, 1944), p. 215.

2. See, for example, *JAMA*, 1 November 1947; *Lancet*, 22 November 1947.

Descriptive Analysis

1. Ludwik Fleck, "Towards a Free and More Human Science," unpublished summary of his views written 1960 in English by Fleck, then in Israel. This article was rejected for publication in 1961, just before Fleck's death, and again in 1978, as having been "overtaken" during the intervening decades.

2. For a similar analysis of the fulfillment of the ancient dream of alchemy, see Thaddeus J. Trenn, "The Justification of Transmutation: Speculations of Ramsay and Experiments of Rutherford," *Ambix* 21 (1974): 53–77.

3. The standard objections to self-justification or self-verification of theories do not pertain, since the "truth" of his position is dependent not upon any philosophical analysis of statements but rather upon historical and sociological development.

4. For recent work in the area of scientific cognition, see David Bloor, *Knowledge and Social Imagery* (London, 1976), and Larry Laudan, *Progress and Its Problems: Towards a Theory of Scientific Growth* (London, 1977), esp. chap. 7.

For empirical data on the formation of scientific specialties, see Henry G. Small, "A Co-Citation Model of a Scientific Specialty: A Longitudinal Study of Colagen Research," *Social Studies of Science* 7 (1977): 139–66; "Structural Dynamics of Scientific Literature," *International Classification* 3 (1976): 67–74. See also E. Garfield, M. V. Malin, and H. G. Small, "Citation Data as Science Indicators," in *Toward a Metric of Science*, ed. Y. Elkana, J. Lederberg, R. K. Merton, A. Thackray, and H. Zuckerman (New York, 1977).

Bibliography

Almenar, Joannis de. *De morbo Gallico libellus*. Venice, 1502. In Luisinus, *Aphrodisiacus* (q.v.), pp. 359–70.

Baas, Johann H. *Grundriss der Geschichte der Medizin und des heilenden Standes*. Stuttgart: Enke, 1876. (Trans.: *Outlines of the History of Medicine and the Medical Profession*. New York, 1889.)

Bartholin, Thomas. *Anatome ex omnium veterum recentiorumque observationibus*. Leyden, 1673.

Berengarius da Carpi, Giacomo. *Commentaria super anatomia Mundini*. Bologna, 1521.

———. *Isagogae breves*. Bologna, 1522.

Bethe, Albrecht. "Kritische Betrachtungen über den vorklinischen Unterricht." *Klinische Wochenschrift* 7 (1928): 1481–83.

———. "Form und Geschehen im Denken des heutigen Arztes." *Klinische Wochenschrift* 7 (1928): 2402–5.

Bierkowski, Ludwig J. S. von. *Choroby syfilityczne czyli weneryczne oraz sposoby ich leczenia*. Krakow: Gieszkowskiego, 1833.

Bloch, Iwan. *Der Ursprung der Syphilis: Eine medizinische und kulturgeschichtliche Untersuchung*. 2 vols. in one. Jena: Fischer, 1901–11.

Bölsche, Wilhelm. *Ernst Haeckel: Ein Lebensbild*. Leipzig, 1900. 2d ed., Berlin: Seeman, 1905. (Volksausgabe, Berlin: Bondi, 1907.)

Bordet, Jules, and Gengou, Octave. "Sur l'existence de substances sensibilisatrices dans la plupart des sérums antimicrobiens." *Annales de l'Institut Pasteur* 15 (1901): 289–302.

Borgarutius, Prosperus. *De morbo Gallico, methodus.* In Luisinus, *De morbo Gallico* (q.v.), vol. 2, pp. 150-85; Luisinus, *Aphrodisiacus* (q.v.), pp. 1117-54.

Brassavolus, Antonius Musa. *De morbo Gallico, liber.* In Luisinus, *De morbo Gallico* (q.v.), vol. 1, pp. 564-610; Luisinus, *Aphrodisiacus* (q.v.), pp. 657-706.

Bruck, Carl. "Zur Geschichte der Serodiagnose der Syphilis." *Berliner klinische Wochenschrift* 58 (1921): 580-81.

————. *Handbuch der Serodiagnose der Syphilis.* 2d ed. Berlin: Springer, 1924.

Carnap, Rudolf. *Der logische Aufbau der Welt.* Berlin, 1928.

————. "Die physikalische Sprache als Universalsprache der Wissenschaft." *Erkenntnis* 2 (1931): 432-65.

————. "Ueber Protokollsätze." *Erkenntnis* 3 (1932): 215-28.

Cataneus de Lacumarcino, Jacobus. *Tractatus de morbo Gallico.* Taurini: Silva, 1532. In Luisinus, *Aphrodisiacus,* pp. 139-68.

Citron, Julius. *Die Methoden der Immunodiagnostik und Immunotherapie und ihre praktische Verwertung.* Leipzig: Thieme, 1910.

Fick, Rudolf. "Betrachtungen über den vorklinischen Unterricht." *Klinische Wochenschrift* 7 (1928): 1921-23.

Fleck, Ludwik, and Elster, Olga. "Zur Variabilität der Streptokokken." *Zentralblatt für Bakteriologie,* Abt. 1, 125 (1932): 180-200.

Flügge, Carl. *Die Mikroorganismen. Mit besonderer Berücksichtigung der Ätiologie der Infektionskrankheiten.* Leipzig: Vogel, 2d ed. 1886; 3d ed. 1896.

Fontanus, Nicolaus, ed. *Librorum Andreas Vesalius de humani corporis fabrica epitome.* Cum annotationibus Nicolai Fontani. Amsterdam, 1642.

Fracanzanus, Antonius. *De morbo Gallico fragmenta quaedam elegantissima ex lectionibus anni 1563.* Padua, 1563.

Fracastoro, Girolamo. *Syphilis, sive morbi Gallici.* Verona, 1530. In Luisinus, *Aphrodisiacus* (q.v.), pp. 183-98.

Freud, Sigmund. *Massenpsychologie und Ich-Analyse.* Leipzig: Internationaler Psychoanalytischer Verlag, 1921.

Frizimelica, Franciscus. *De morbo Gallico, tractatus.* In Luisinus, *De morbo Gallico* (q.v.), vol. 2, pp. 28-43; Luisinus, *Aphrodisiacus* (q.v.), pp. 985-1000.

Fröhlich, Friedrich W. "Ueber den vorklinischen Unterricht." *Klinische Wochenschrift* 7 (1928): 1923-24.

Fuchs, C. H. *Die ältesten Schriftsteller über die Lustseuche in Deutschland von 1495-1510.* Göttingen: Dieterich, 1843.

Gärtner, Carl Friedrich von. *Observata quaedam circa urinae naturam.*

Tübingen, 1796. (Trans.: "Ueber Harn." *Reil Archiv für Physiologie* 2 [1797]: 169-203.)

Geigel, Alois. *Geschichte, Pathologie und Therapie der Syphilis.* Würzburg: Stuber, 1867.

Goldstein, Kurt. "Betrachtungen über den vorklinischen Unterricht." *Klinische Wochenschrift* 7 (1928): 2399-402.

Göppert, Ernst. "Kritische Betrachtungen über den vorklinischen Unterricht." *Klinische Wochenschrift* 7 (1928): 1876.

Gottstein, Adolf. *Die Lehre von den Epidemien.* Berlin: Springer, 1929.

Gradmann, Hans. "Die harmonische Lebenseinheit vom Standpunkt exakter Naturwissenschaft." *Naturwissenschaften* 18 (1930): 641-44, 662-66.

Gumplowicz, Ludwig. *Grundriss der Soziologie.* Vienna, 1885; 2d ed., 1905.

Haeckel, Ernst. *Natürliche Schöpfungs-Geschichte.* Berlin, 1868.

Heitzmann, Carl. *Die descriptive und topographische Anatomie des Menschen.* 5th ed. Vienna: Braumüller, 1888.

Hergt, F. C. *Geschichte, Erkenntnis und Heilung der Lustseuche.* Hadamar, 1826.

Hermann, Josef. *Es gibt keine constitutionelle Syphilis: Ein Trostwort für die gesamte Menschheit.* Hagen in Westphalia, 1891; 4th ed., Leipzig: Otto, 1903.

Hirszfeld, Ludwig. "Prolegomena zur Immunitätslehre." *Klinische Wochenschrift* 10 (1931): 2153-59.

Hoffmann, Erich. *See* Schaudinn, Fritz R., and Hoffmann, Erich.

Hohenheim, Theophrastus von. *See* Paracelsus.

Huser, J. *See* Paracelsus.

Jerusalem, Wilhelm. "Die soziologische Bedingtheit des Denkens und der Denkformen." In *Versuche zu einer Soziologie des Wissens,* pp. 182-207, ed. Max Scheler. Leipzig and Munich: Duncker und Humblot, 1924.

———, ed. *See* Lévy-Bruhl, Lucien, *Das Denken der Naturvölker.*

Kant, Immanuel. *Kritik der reinen Vernunft,* "Vorrede zur zweiten Auflage." In *Sämtliche Werke,* vol. 3. Leipzig: Insel, 1921-22.

Kelsen, Hans. "Der Begriff des Staates und die Sozialpsychologie." *Imago* 8 (1922): 97-141.

Kirchberger, Paul. *Die Entwicklung der Atomtheorie.* Karlsruhe: Müller, 1922; 2d ed., 1929.

Koch, Richard, and Rosenstock, Eugen, eds. *Paracelsus: Krankheit und Glaube (Fünf Bücher über die unsichtbaren Krankheiten).* Stuttgart: Frommann, 1923.

Kolle, Wilhelm; Kraus, Rudolf; and Uhlenhuth, Paul. *Handbuch der pathogenen Mikroorganismen.* 3d ed., vol. 7. Jena: Fischer, and Berlin-Vienna: Urban und Schwarzenberg, 1930.

Lange, Friedrich A. *Geschichte des Materialismus und Kritik seiner Bedeutung in der Gegenwart.* 1866. Leipzig: Reclam, 1905. (Trans.: *History of Materialism and Criticism of Its Present Importance.* 3d ed. London: Routledge and Kegan Paul, 1925.)

Laubenheimer, Kurt. "Serumdiagnose der Syphilis." In Kolle, Kraus, and Uhlenhuth (q.v.), vol. 7, pp. 216–336. (Completed in 1927.)

LeBon, Gustave. *La psychologie des foules.* Paris, 1895. (Trans.: *Psychologie der Massen.* 2d ed. Leipzig: Kröner, 1912.)

Lehmann, Karl Bernhard, and Neumann, Rudolf Otto. *Atlas und Grundriss der Bakteriologie und Lehrbuch der speziellen bakteriologischen Diagnostik.* Munich, 1896; 2d ed., 1899; 7th ed., 1926–27.

Lesky, E. "Von Schmier- und Räucherkuren zur modernen Syphilis-therapie." *Ciba-Zeitschrift* 8 (1959): 3174–89.

Lévy-Bruhl, Lucien. *Das Denken der Naturvölker.* Ed. Wilhelm Jerusalem. 2d ed. Vienna and Leipzig: Braumüller, 1926.

Lostorfer, Adolf. "Ueber die Möglichkeit der Diagnose der Syphilis mittelst der mikroskopischen Blutuntersuchung." *Medizinische Jahrbücher (Gesellschaft der Aerzte,* Vienna), 1872, pp. 96–105.

Löw, Joseph. *Ueber den Urin als diagnostisches und prognostisches Zeichen in physiologischer und pathologischer Hinsicht.* Landshut: Thomann, 1809; 2d ed., 1815.

Luisinus, Aloysius, ed. *De morbo Gallico omnia quae extant...* 2 vols. Venice: Zilettus, 1566–67.

―――. *Aphrodisiacus, sive de lue venerea, vel morbo Gallico Opus.* Leyden: Langerak & Verbeek, 1728. (The Boerhaave edition.)

Mach, Ernst. *Die Mechanik in ihrer Entwicklung.* Leipzig, 1833; 6th ed., 1908.

Marx, Ernst. *Die experimentelle Diagnostik, Serumtherapie und Prophylaxe der Infektionskrankheiten.* Berlin, 1902.

McDougall, William. *The Group Mind: A Sketch of the Principles of Collective Psychology, with Some Attempt to Apply Them to the Interpretation of National Life and Character.* Cambridge: Cambridge University Press, 1920.

Metzger, Wolfgang. "Psychologie Mitteilungen: Laut und Sinn." *Naturwissenschaften* 17 (1929): 846.

Meyer-Steineg, Theodor, and Sudhoff, Karl. *Geschichte der Medizin im Überblick mit Abbildungen.* Jena: Fischer, 1921. 3d ed., 1928.

Möller, Johannes, and Müller, Paul. *Grundriss der Anatomie des Menschen für Studium und Praxis.* 2d ed. Leipzig: Veit, 1914.

Montagnana, Bartholomeus. *De morbo Gallico, consilium.* In Luisinus, *De morbo Gallico* (q.v.), vol. 2, pp. 1-8; Luisinus, *Aphrodisiacus* (q.v.), pp. 957-66.

Nägeli, Otto. *Allgemeine Konstitutionslehre in naturwissenschaftlicher und medizinischer Betrachtung.* Berlin: Springer, 1927.

Nagelschmidt, Franz. *Ueber Immunität bei Syphilis nebst Bemerkungen über Diagnostik und Serotherapie der Syphilis.* Berlin: Hirschwald, 1904.

Nauwerck, Coelestin. *Sektionstechnik für Studierende und Aerzte.* 5th ed. Jena, 1912.

Ostwald, Wilhelm, *Jak powstała Chemja.* 1906. (Translated from *Leitlinien der Chemie: Sieben gemeinverständliche Vorträge aus der Geschichte der Chemie.* Leipzig: Akademische Verlagsgesellschaft, 1906.)

Paracelsus (Theophrastus von Hohenheim). *Bücher und Schriften.* 10 vols. Ed. Johannes Huser. Basel: Waldkirch, 1589-91. (Reprint, with foreword by Kurt Goldammer, 6 vols. Hildesheim: Olms, 1971-77.)

————. *See* Koch, Richard, and Rosenstock, Eugen, *Paracelsus.*

Petersen, Hans. "Ueber die Rolle der Anatomie im Lehrgang des künftigen Arztes." *Klinische Wochenschrift* 7 (1928): 1872-75.

Plaut, Felix. "Die theoretische Begründung der Wassermannschen Reaktion." *Münchener medizinische Wochenschrift* 78 (1931): 1461-63.

Popper, Karl. *Logik der Forschung.* Vienna, 1935. (Trans.: *The Logic of Scientific Discovery.* London: Hutchinson, 1959.)

Proksch, J. K. *Die Litteratur über die venerischen Krankheiten.* 5 vols. 1889-1900. (Reprint, 3 vols. Nieuwkoop: De Graaf, 1966.) (Contains many references indicated but not explicitly cited by Fleck.)

Ramsay, William. *Vergangenes und Künftiges aus der Chemie: Biographische und chemische Essays.* 2d ed. Leipzig, 1913. (Translated from *Essays, Biographical and Chemical.* London: Constable, 1908.)

Reich, Eduard. *Ueber den Einfluss der Syphilis auf das Familienleben.* Amsterdam: Dieckmann, 1887; 2d ed., 1894.

Roth, Moritz. *Andreas Vesalius Bruxellensis.* Berlin: Reimer, 1892.

Rinius, Benedictus. *De morbo Gallico, tractatus.* In Luisinus, *De morbo Gallico* (q.v.), vol. 2, pp. 14-27; Luisinus, *Aphrodisiacus* (q.v.), pp. 971-84.

Schaudinn, Fritz R., and Hoffman, Erich. "Vorläufiger Bericht über das Vorkommen von Spirochaeten in syphilitischen Krankheitsprodukten und bei Papillomen." *Arbeiten aus dem Kaiserlichen Gesundheitsamte* 22 (1905): 527-34. Also in *Vorträge und Urkunden zur 25-jährigen Wiederkehr der Entdeckung des Syphiliserregers (Spirochaeta pallida),* ed. Erich Hoffman. Berlin: Karger, 1930.

Scheler, Max, ed. *Versuche zu einer Soziologie des Wissens.* Munich and Leipzig: Duncker und Humblot, 1924.

Schreger, Odilo. *Studiosus jovialis; seu, Auxilia ad jocose et honeste discurrendum, in gratiam et usum studiosorum juvenum, aliorumque litteratorum virorum, honestae recreationis amantium collecta.* Munich: Gastl, 1749.

Schuberg, August, and Schlossberger, Hans. "Zum 25. Jahrestag der Entdeckung der Spirochaete pallida." *Klinische Wochenschrift* 9 (1930): 582–86.

Siegel, John. "Untersuchungen über die Aetiologie der Syphilis." *Abhandlungen der königlich preussischen Akademie der Wissenschaften* (Berlin), *Anhang,* 1905, no. 3, pp. 1–15.

Simmel, Georg. *Soziologie: Untersuchungen über die Formen der Vergesellschaftung.* Munich and Leipzig: Duncker und Humblot, 1908.

Simon, Friedrich Alexander. *Ricord's Lehre von der Syphilis, ihre bedenklichen Mängel und groben Irrthümer kritisch beleuchtet und durch zahlreiche, schwierige und verzweifelte Krankheitsfälle erläutert; ein praktisches Handbuch über Syphilis.* Hamburg: Hoffman und Campe, 1851–52.

Sudhoff, Karl. *Tradition und Naturbeobachtung in den Illustrationen medizinischer Handschriften und Frühdrucke vornehmlich des 15. Jahrhunderts.* Studien zur Geschichte der Medizin, vol. 1. Leipzig: Barth, 1907.

———. *Der Ursprung der Syphilis.* Leipzig: Vogel, 1913.

———. *Kurzes Handbuch der Geschichte der Medizin.* Berlin: Karger, 1922.

Sydenham, Thomas. *Opera omnia medica.* Venice, 1735.

Toldt, Carl. *Anatomischer Atlas für Studierende und Aerzte,* 3 vols. Berlin and Vienna: Urban und Schwarzenberg, 1900–1903.

Tomitanus, Bernardus. *De morbo Gallico, libri duo.* In Luisinus, *De morbo Gallico* (q.v.), vol. 2, pp. 58–139; Luisinus, *Aphrodisiacus* (q.v.), pp. 1015–1106.

Uexküll, Jakob von. *Theoretische Biologie.* 2d ed. Berlin: Springer, 1928.

Vesalius, Andreas. *De Humani corporis fabrica librorum, Epitome.* Basel: Oporinus, 1543.

Wassermann, August von. "Neue experimentelle Forschungen über Syphilis." *Berliner klinische Wochenschrift* 58 (1921): 193–97.

———. "Zur Geschichte der Serodiagnostik der Syphilis." *Berliner klinische Wochenschrift* 58 (1921): 1194–95.

Wassermann, August von; Neisser, Albert; and Bruck, Carl. "Eine serodiagnostische Reaktion bei Syphilis." *Deutsche medizinische Wochenschrift* 32 (1906): 745-46.

Wassermann, August von; Neisser, Albert; Bruck, Carl; and Schucht, A. "Weitere Mitteilungen über den Nachweis spezifisch luetischer Substanzen durch Komplementverankerung." *Zeitschrift für Hygiene und Infektionskrankheiten* 55 (1906): 451-77.

Weil, Edmund. "Das Problem der Serologie der Lues in der Darstellung Wassermanns." *Berliner klinische Wochenschrift* 58 (1921): 966-70.

Weindler, Fritz. *Geschichte der gynäkologisch-anatomischen Abbildung.* Dresden: Zahn und Jaensch, 1908.

Wendt, Johann. *Die Lustseuche in allen ihren Richtungen und in allen ihren Gestalten.* Breslau: Korn, 1816. 3d ed., Vienna, 1827.

Widman, Johann. *Tractatus de pustulis que morbo qui vulgato nomine dicunter mal de Franzos.* Tübingen, 1497. In C. H. Fuchs, *Die ältesten Schriftsteller* (q.v.), pp. 95-112.

Wiener, Emmerich, and von Torday, Árpád. "Eigenartig spezifisches Verhalten luetischer und karzinomatöser Sera gegen bestimmte Chemikalien." *Deutsche medizinische Wochenschrift* 40 (1914): 429-30.

Wood, John George. *Homes without Hands: Being a Description of the Habitations of Animals, Classed According to their Principle of Construction.* New York and London, 1866.

Wundt, Wilhelm Max. *Logik: Eine Untersuchung der Prinzipien der Erkenntnis und der Methoden wissenschaftlicher Forschung.* Stuttgart: Enke, 1893-95.

Early Reviews *of Entstehung und Entwicklung einer wissenschaftlichen Tatsache*

Anon. *Baslersche Nachrichten,* no. 29, 25 July 1937 (signed "H.C.").

Anon. *Natur und Kultur* 35 (1938): 143-44.

Anon. *Wiener klinische Wochenschrift* 49 (1936): 1470.

Baege, Max H. *Natur und Geist* 5 (1937): 380-81.

Bing, Robert. *Schweizerische medizinische Wochenschrift* 66 (1936): 303.

Caspari, Wilhelm. *Umschau* 40 (1936): 338.

Guérard des Lauriers, M. L. *Revue des sciences philosophiques et théologiques* 26 (1937): 320-21.

Fischer, Franz. *Nervenarzt* 9 (1936): 137-38.

Haeberlin, Carl. *Deutsche medizinische Wochenschrift* 63 (1937): 244.

Kroh, Oswald. *Zeitschrift für Psychologie* 138 (1936): 163-64.

Petersen, Hans. *Klinische Wochenschrift* 15 (1936): 239-42.

Index

historical   vs.   epistemological

does it explain
historical changes
in scientific thought?

what's the unit of analysis —
    a "fact" ?
                             "web" ⟨ a person?
    a "thought style" ? ⟩ of ⟨
     an "idea" ?   ideas   a collective?
    (proto-idea?)   all
                      defining
                      each other